The Great Chain of History

Kayw

BUCKLAND WILLIAM #
1784-1856

GEOLOGY

HISTORY

4780 ?

No. 6092 294 07

# THE GREAT CHAIN OF HISTORY

*William Buckland*
*and the English School of Geology*
*(1814–1849)*

NICOLAAS A. RUPKE

CLARENDON PRESS · OXFORD
1983

Oxford University Press, Walton Street, Oxford OX2 6DP
London Glasgow New York Toronto
Delhi Bombay Calcutta Madras Karachi
Kuala Lumpur Singapore Hong Kong Tokyo
Nairobi Dar es Salaam Cape Town
Melbourne Auckland
and associated companies in
Beirut Berlin Ibadan Mexico City Nicosia

Oxford is a trade mark of Oxford University Press

Published in the United States
by Oxford University Press, New York

British Library Cataloguing in Publication Data

Rupke, Nicolaas A.
The great chain of history.
1. Historical geology        2. Geologists
—England—History—19th century
I. Title
551.7'001        QE28.3
ISBN 0-19-822907-0

Typeset by Cotswold Typesetting Ltd, Cheltenham
Printed in Great Britain
at the University Press, Oxford
by Eric Buckley
Printer to the University

*For my Friends*

# Preface

This study is intended as a contribution to the cultural history of early nineteenth-century England. Attention is focused on the 1820s and 1830s when geology was at the cutting edge of science. England's most accomplished and popular geologist of the period was William Buckland. He, and a number of like-minded colleagues, formed a loosely-knit network, the 'English school', to which we owe some of the most outstanding discoveries in paleontology and stratigraphy. The discussion of the English school is based on published sources and much previously unknown manuscript material.

Buckland and his circle belonged to Oxford and Cambridge, and this institutional connection influenced the content of their geology. Conversely, geology figured prominently in issues of university reform. It reduced the authority of theology, undermining both Mosaical cosmogony and New Testament eschatology. Moreover, in opposition to it, a biblicist sub-culture formed, which today is known as fundamentalism. Selected examples of the impact of geology on the visual arts (John Martin), on literature (Tennyson), moral philosophy, and such instances of social opinion as the idea of progress, are discussed.

This study was made while the author held research fellowships at, respectively, St. Peter's College and Wolfson College, Oxford. Debt is acknowledged to the fellows of both colleges for many stimulating discussions, and especially to Sir Alec Cairncross and Sir Henry Fisher for their sustained interest and support. Professor Margaret Gowing has shown generous interest from the beginning, and her warm support and careful reading of the manuscript have been invaluable. Much encouragement has been received from Dr Alistair Crombie; Mr Rom Harré; and Mr Wyllie Wright, who also scrutinized the manuscript.

Generous assistance has been granted at the Department of Geology and the University Museum, Oxford, by Professor E. A. Vincent, by Dr Jim Kennedy, who made all the Buckland papers freely available, and by Mrs Ballard, Phil Powell, and Mrs

Whittaker. Similar co-operation from the various libraries and archives, named in the list of manuscript sources, is gratefully acknowledged.

Much profit and pleasure was gained by regular discussions with my friends, in the King's Arms or in other locales within striking distance of the Bodleian Library, especially with Drs Moti Feingold, Jeremy George, Willem Hackmann, Michael Lockwood, Renato Mazzolini, Oliver Nicholson, Kevin Sharp, Janis Spurlock, and, last but not least, Francis Warner. Michael, in addition, improved the text.

Financial assistance at the outset of this study was generously given by the Howard Foundation at Brown University. The manuscript received its finishing touches while the author enjoyed the support of the Humboldt Stiftung at the University of Tübingen. Professor Wolf Freiherr von Engelhardt freely shared his rich store of knowledge about the geology of the Goethe era. Mrs Cremer meticulously typed the manuscript.

Summer 1982
University Museum
Oxford

# Contents

ix

# Illustrations

# Citation

Original passages are quoted in order to convey contemporary flavour and to give this text the function of a source book of English historical geology. The spelling mistakes in quotations from unpublished sources have been corrected and capitals and punctuation adjusted to the OUP house style. German and French sources are quoted in the original except when an English translation was widely available at the same time. Italics in quotations are invariably original. In the text the delivery date of lectures and sermons is indicated; in the bibliography the date of publication is given. Dates given for books in the bibliography refer to the particular editions used for this study; these do not always tally with the dates in the text where, unless otherwise stated, the first editions are indicated. Footnotes contain abbreviated reference material, more fully documented in the bibliography. Mention of the Geological and Royal Societies refers to those of London. Familiar paleontological names are given in roman type, while less familiar ones have been italicized in accordance with taxonomic practice. The German *umlaut*, when written as a diaeresis, has not been considered in determining the alphabetical order of bibliography and index.

# INTRODUCTION

For some years past, these newly created sciences
(Geology and Mineralogy) have formed a leading
subject of education in most Universities on the continent, and a competent knowledge of them is now
possessed by the majority of intelligent persons in our
own country; and though it might on no account be
desirable to surrender a single particle of our own
peculiar, and, as we think, better system of Classical
Education, there seems to be no necessity for making
that system an exclusive one; nor can any evil be anticipated from their being admitted to serve at least a subordinate ministry in the temple of our Academical
Institutions.

Buckland, *Vindiciae Geologicae*, 3

# 1
# Geology at Oxford and the English School

## THE NEW EARTH HISTORY

During the late eighteenth and early nineteenth centuries geology opened up a vast and unfamiliar perspective of earth history. The study of rocks and fossils showed that the history of the earth had not covered the same stretch of time as the history of mankind, but extended back immeasurably before the appearance of man. The individual discoveries of geology, especially in paleontology, were exciting in isolation. Even more significant was the conclusion that pre-human earth history had not been a single period of continuity, but a concatenation of successive worlds, i.e. of periods of geological history each characterized by a particular extinct flora and fauna. Moreover, it appeared that the nature of the historical succession had been progressive; that successive worlds increasingly resembled our present world, both with respect to its inhabitants and to the environmental conditions under which they lived.

Thus a great chain of history appeared to unroll from the new understanding of rocks and fossils, analogous and complementary to the great chain of being known in contemporary natural history. This new perspective of earth history equalled the Copernican Revolution in its intellectual implications; it reduced the relative significance of the human world in time, just as early modern astronomy had diminished it in space.

The reconstruction of pre-human periods of earth history was based on two separate areas of study. The first was allied to chemistry and consisted of the mineralogical classification of rock formations (petrography in today's terminology). It was institutionalized at the Mining Academy of Freiberg where Abraham Gottlob Werner developed a geological periodization based on the composition of rocks.[1] The second area of study was allied to medicine; using comparative anatomy a number of physicians and

[1] *Kurze Klassifikation und Beschreibung der verschiedenen Gebirgsarten*, 1787.

surgeons recognized that large fossil bones belonged to species no longer alive today. Johann Friedrich Blumenbach, professor of medicine at the University of Göttingen, concluded from his study of fossils that a pre-Adamite world had existed.[2] George Cuvier made the subject of vertebrate paleontology his particular domain, and the Jardin du Roi in Paris became after Freiberg a major centre of historical geology.[3]

Already Buffon's *Époques de la nature* (1778) had made a large and international readership familiar with the notion of pre-human periods of earth history. This change in the perception of the past was part of a wider new philosophy of history. R. G. Collingwood has described how the idea of history underwent a major change during this period, exemplified by Herder's *Ideen zur Philosophie der Geschichte der Menschheit* (1784–91).[4] The first volume discussed the relationship of the earth to other celestial bodies, its successive stages of pre-human history, the progressive appearance of life on earth, and the place of man in nature.

Many of the most outstanding figures of the period were fascinated by earth science because of its historicity. Goethe was intensely interested in both fossils and minerals. Novalis (Friedrich von Hardenberg) actually enrolled as a student at Freiberg's Mining Academy.[5] Cuvier himself discussed the relationship of earth history with human history in the popular preliminary discourse to his *Ossemens fossiles* (1811).

Various historians have discussed the change from a static natural history to a temporalized history of nature.[6] Their studies make little mention of the English contributions to the new earth history. Admittedly, the late-eighteenth-century phase of historical geology effectively bypassed England, even though Jean André Deluc, who resided for much of his life in London, created some awareness of a pre-Adamite earth history. Early in the nineteenth century Humphry Davy lectured on the new geology at the Royal Institution. An empirical and rather isolated provincial tradition of stratigraphy existed with William Smith.

---

[2] *Beyträge zur Naturgeschichte,* i (1790).
[3] See Rudwick, *Meaning of Fossils,* 1972, 101 ff.
[4] *The Idea of History,* 1946, part iii.
[5] *Goethe, Schriften zur Naturwissenschaft,* ed. Wolf *et al.,* xi (1970). See Schulz, 'Novalis und der Bergbau', *Freiberger Forschungshefte,* D 11 (1955), 242–55.
[6] E.g. Foucault, *Les Mots et les choses,* 1966; Lepenies, *Das Ende der Naturgeschichte,* 1976; Engelhardt, *Historisches Bewusstsein in der Naturwissenschaft,* 1979.

During the second decade of the nineteenth century, however, a school of historical geology began to form at the University of Oxford. Its motive forces were William Buckland and a circle of colleagues among whom William Daniel Conybeare was the most outstanding figure. This circle grew to encompass in the course of the 1820s Adam Sedgwick and other Cambridge men such as William Whewell (figures 1–4). At the same time it drew into its orbit various metropolitan and provincial devotees of the new science. These men, interested in paleontology and stratigraphy, represented by about 1830 a characteristically English school of historical geology.

Its institutional identity did not exist in a society nor in a journal, but centred on the geology lecture courses at Oxford and Cambridge, England's ancient universities. In contrast to continental geology the English school remained shielded from the increasingly secular philosophy of history propagated by Kant, Hegel, and others. It adjusted itself to the indigenous tradition of Anglican learning in which natural and revealed religion were intimately interwoven with the form and substance of science. This adjustment gave to the English school some of its distinctive features and was the cause of a number of its successes in, for example, the functional anatomy of fossils.

The modern perspective of earth history owes little to Scottish geology. Huttonian uniformitarianism was fundamentally uncongenial to historical geology. Its vision of a permanent present made James Hutton object to invertebrate extinction. It also caused Charles Lyell to reject the evidence for the progressive succession of fossils. The common notion that modern geology originated with uniformitarianism[7] is a hindrance to the unencumbered study of the origin of the new geology. The distinctive nature of English geology was in fact accentuated by the conscious manner in which it set itself apart from the Scottish tradition.

## BUCKLAND AND HIS CIRCLE

*Scientific biography*

Buckland came from a Devonshire family of Anglican clerics. He was born in 1784, the year in which the first volume of Herder's

---

[7] E.g. Haber, *The Age of the World*, 1959; Bailey, *James Hutton, the Founder of Modern Geology*, 1967; Wilson, *Charles Lyell: the Years to 1841*, 1972.

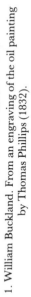

2. Adam Sedgwick. From an engraving of the oil painting by Thomas Phillips (1832).

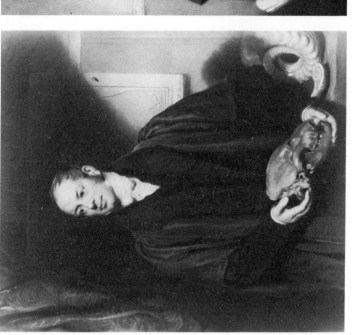

1. William Buckland. From an engraving of the oil painting by Thomas Phillips (1832).

*Ideen* was published. In 1801 he won a scholarship to Corpus Christi College, Oxford, just when new examination statutes began to take effect, making university study a more serious affair than it had been during much of the previous century. In due course he graduated (1805), took holy orders, and was elected a fellow of his college (1808).

Buckland's formal education focused on classics, but his hobby was science. He took time to attend the lectures by Christopher Pegge, the regius professor of medicine, and by John Kidd, Aldrichian professor of chemistry and later successor to Pegge. Buckland's enthusiasm and real talent earned him the readership of mineralogy (1813), in succession to Kidd, and a few years later a readership of geology was created (1818) for the purpose of adding to Buckland's meagre income.[8]

An annual lecture course on geology was given by Buckland from 1814 till 1849, when a mental disability forced him to retire. Attendance at his lectures was not compulsory, but both junior and senior members of the university crowded his lecture theatre, especially during the early half of his Oxford career. Their popularity was not only based on the quality and novelty of Buckland's work, but also on his oratory and ability to amuse or shock an audience. His prose, corrected by his wife Mary Morland, was very fine indeed, and probably superior to that of any of his colleagues. However, his sense of humour was rather coarse, and recalled an earlier age; to earnest Victorians such as Charles Darwin the lecturing antics of Buckland, which also enlivened many a debate at the Geological Society, were undignified buffoonery.

In the early 1820s, at the age of thirty-six, Buckland developed a theory which made him England's leading geologist and probably the most talked-of scientist of the decade—a man to be mentioned in the same breath as Cuvier. His theory interpreted a number of English bone caves as antediluvian dens of hyenas, the most famous of which was Kirkdale Cave in Yorkshire. Buckland's cave paleontology became an enduring cornerstone of his subject. It represented the first major ecological study of fossils, examined not just as taxonomic entities, but as members of fossil communities, to be interpreted with the aid of present-day analogues.

[8] Gordon, *Life and Correspondence of Buckland,* 1894; Edmonds, 'Founding of the Oxford Readership in Geology', *Notes and Records Roy. Soc.* xxxiv (1979), 33–51.

4. William Whewell. From an engraving by E. U. Eddis (1835).

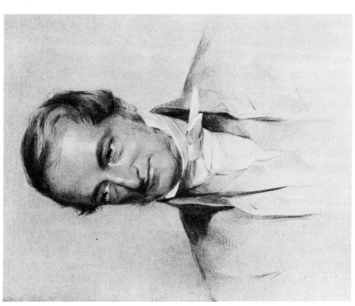

3. William Daniel Conybeare. From an engraving of an oil painting, aged sixty-five.

Buckland integrated his cave paleontology with diluvialism, which attributed a variety of geological surface phenomena to the waters of the biblical deluge. Diluvial geology was formulated in Buckland's inaugural *Vindiciae Geologicae* (1819) and his *Reliquiae Diluvianae* (1823). It represented the main focus of geological theory during the 1820s, the diluvial decade, and provided a frame of reference for observations by many of his colleagues.

By about 1830 a new and wider synthesis replaced the diluvialism of the 1820s. The geological annals of the antediluvial period and of the diluvial event consisted merely of superficial deposits, mud layers in caves and gravel strewn across the countryside. By implication, the very thick and extensive rock formations on which the diluvial deposits rested, and which contained very different fossil species from the ones in the mud and gravel, had to represent periods of geological history stretching back far beyond even the antediluvial epoch of the cave hyenas.

A major part of Buckland's career and those of his colleagues was taken up with the task of unravelling this lengthy and complex geological history: attempting to establish the exact sequence of rock formations through time, to identify and reconstruct the fossils in these rocks, to speculate about the apparently progressive nature of earth history, and to reconcile the new perspective of geological time with the traditional view of biblical history. Buckland's Bridgewater Treatise (see ch. 18) on *Geology and Mineralogy* (1836) was the most complete summary of the geology of the English school and of its progressivist synthesis of earth history. Although not as compendious as the work of some continental colleagues, it was of considerably higher quality because of its use of functional morphology, effectively missing in Georg August Goldfuss's *Petrefacta Germanica* (1826–44) or Heinrich Georg Bronn's *Lethaea Geognostica* (1835–8).

At Oxford Buckland befriended men of similar background and scientific interests. Among these were the brothers Conybeare, John Josiah and William Daniel. The latter, born in 1787, came up to Oxford at the same time as Robert Peel and Richard Whately. Regrettably, he left Oxford to marry, retiring to a country curacy in 1814. But he kept in touch with his *alma mater* and co-operated with Buckland on a variety of geological projects. In addition, Conybeare added significantly to the reputation of Oxford geology by writing a textbook (1822), an enlarged edition of the *Outline of the Geology of England and Wales* (1818) by William Phillips.[9]

Although Buckland's cleverness in geology was unsurpassed, Conybeare was the more impressive intellectually. He acted with all the self-confidence of a man who had obtained a first in classics at Christ Church and who lived in the comfortable embrace of the ecclesiastical establishment. Buckland excelled in the detail of geological argument, but he needed the support of Conybeare in matters of theory. Buckland's emotional constitution required the applause of his audience. Conybeare was perfectly content laughing at his own jokes. When under attack from the Scottish school, Buckland relied on Conybeare to defend him.

At Cambridge the new earth history was taught by Adam Sedgwick, appointed to the Woodwardian chair in 1818. By his own account he knew little or nothing about geology at the time of his professorial appointment, but he took to his new task with enthusiasm, guided by Conybeare. Among his early publications was a defence of Buckland's diluvialism which helped to establish the distinction between Alluvium and Diluvium. Sedgwick's anniversary addresses to the Geological Society (1830, 1831) and his famous sermon at Trinity College (1832), published and several times reprinted under the title *A Discourse on the Studies of the University,* contributed significantly to the definition of English geology and of its progressivist synthesis. These contributions firmly established Sedgwick as a thinker in his own right.[10]

Nevertheless, he came second to Buckland in nearly every respect. Sedgwick was born in 1785, Buckland's junior by a year; he taught at England's second university; he began his course of geology lectures half a decade after Buckland had started lecturing at Oxford; he was elected to the presidency of the Geological Society five years after the first time Buckland occupied the chair; he was president of the British Association in succession to his Oxford counterpart; and he was awarded the Royal Society's Copley Medal and the Wollaston Medal by the Geological Society long after Buckland had enjoyed the same honours. However, Sedgwick's lectures were as famous and popular as the geology course at Oxford. His manly countenance was matched by the

[9] North, 'Dean Conybeare, Geologist', *Report and Trans. Cardiff Naturalists' Soc.* lxvi (1933), 15–68.

[10] Clark and Hughes, *Life and Letters of Sedgwick,* 1890.

forcefulness of his pronouncements on matters of controversy. Sedgwick tended to procrastinate and he never wrote a book, but unlike Buckland he did involve himself in university affairs.

Buckland, Conybeare, and Sedgwick belonged to a generation of Oxbridge men who were educated during the insular period of the Napoleonic wars. The Second Peace of Paris in 1815 facilitated international travel and helped transplant historical geology from Germany and France on to English soil. In this Buckland and his circle were aided by several younger men who had been born during the 1790s. They loyally supported Buckland's diluvialism and contributed to the establishment of scientific societies, namely the Cambridge Philosophical Society (1819) and its younger Oxford sister, the Ashmolean Society (1828). To this younger generation belonged Charles Giles Bridle Daubeny, who took over from Kidd as Oxford's professor of chemistry (1822), and at Cambridge John Stevens Henslow, who became professor of mineralogy (1822–8). He was succeeded by William Whewell (1828–32), whose book reviews in the *British Critic* and Bridgewater Treatise on *Astronomy and General Physics* (1833) contributed to the definition of the progressivist synthesis of the English school, adding to Sedgwick's work.[11]

Outsiders who were attracted to the new geology included Henry Thomas de la Beche, Roderick Impey Murchison, and John Phillips, who later succeeded Buckland as professor of geology at Oxford. Among the overseas advocates of English geology were the outspoken explorer George William Featherstonhaugh and Benjamin Silliman, who was editor of the *American Journal of Science and Arts*. Buckland's early pupils included Philip de Malpas Grey-Egerton, famous for his collection of fossil fishes, and William Fox-Strangways, who initiated the study of Russian stratigraphy. His sister, Mary Lucy Cole, and her daughters by her first marriage, the Misses Talbot, enthusiastically contributed to Buckland's cave researches from their Glamorganshire base. Charles Lyell was also one of Buckland's early students although in the late 1820s he allied himself increasingly with the Huttonian tradition of his native Scotland.

Part of the inspiration for historical geology came from London,

---

[11] Todhunter, *William Whewell*, 1876; Douglas, *Life and Correspondence of Whewell*, 1881.

where in 1807 the Geological Society had been founded.[12] For some years Buckland and Conybeare formed a relationship of close co-operation with its first president, George Bellas Greenough, with whom they visited both British and continental places of geological interest. Greenough, however, irreparably compromised his reputation with *A Critical Examination of the First Principles of Geology* (1819). Its excessive Baconian empiricism denied the possibility of an integrated earth history in the form of an international strati-graphic table, the very object of Buckland's and Conybeare's interest. Both men bypassed London and associated with Cuvier and his school in Paris, deriving vicarious prestige in the process. Joseph Pentland, assistant to Cuvier at the Jardin du Roi, con-ducted a very valuable middleman correspondence (1820–2) which kept Buckland and Conybeare in touch with Cuvier.[13]

This contact helped Buckland and Conybeare to become recognized authorities of paleontology, until then the domain of physicians and surgeons. Thus England's foremost contributors to the study of fossils during the early part of the nineteenth century were men such as William Clift, Everard Home, and James Parkinson, all pupils of the famous eighteenth-century anatomist and surgeon John Hunter. The latter's collection, which included fossils, was entrusted to the Company of Surgeons in 1799, where it became a focal point of paleontological interest. Clift was appointed curator of the Hunterian Museum. Home succeeded Hunter at St. George's Hospital, and became keeper of Hunter's collection from which a number of papers on anatomy, which he gave to the Royal Society, were allegedly plagiarized. Parkinson, best known for the disease named after him, wrote a monumental paleontology textbook, *Organic Remains of a Former World* (1804–11).[14]

The connection of medicine with paleontology persisted beyond the generation of Hunter's pupils. Gideon Algerton Mantell, a provincial surgeon, was an accomplished vertebrate paleontologist. When Conybeare wrote a paper on reptilian paleontology he apologized for 'intruding on the province of the comparative

[12] See Rudwick, 'Foundation of the Geological Society of London', *Brit. Journ. Hist. Sci.* i (1963), 325–55.

[13] Sarjeant and Delair, 'An Irish Naturalist in Cuvier's Laboratory', *Bull. Brit. Mus. (Nat. Hist.)*, Hist. Ser. vi (1980), 245–319.

[14] See Thackray, 'James Parkinson's *Organic Remains*', *Journ. Soc. Biblphy Nat. Hist.* vii (1976), 451–66.

anatomist'.[15] Richard Owen, England's leading paleontologist of the mid-nineteenth century, was appointed to the Hunterian professorship at the Royal College of Surgeons in 1836. That year Buckland was made an honorary fellow of the Medico-Chirurgical Society of London.

*Politics and religion*

Buckland lived during the so-called 'age of improvement'.[16] Industrial expansion and technological innovation occurred on an unprecedented scale. The railways were developed; steamships shortened voyages across the Channel and the Atlantic; the electric telegraph was put on a commercial basis; the penny post was introduced; and gas lighting became a new source of illumination. The Great Exhibition of 1851 summed up the tremendous material advances made during the first half of the nineteenth century. The same period saw improvement of a social nature. Slavery in the British Empire was abolished; trade unions were organized; the Test Acts were repealed; and the passage of the Catholic Emancipation Act (1829) under Tory government was followed by a variety of Whig reform measures in 1832.[17]

Buckland enthusiastically participated in technological advances, even acting as chairman of the Oxford Gas and Coke Company. Like his close colleagues he supported moderate reform measures, although politically they were not of a single colour. Buckland and Whewell were associated with Peelite Toryism. With Peel Buckland enjoyed a life-long friendship, and the two men corresponded frequently. When Oxford had become unbearable to Buckland, Peel appointed him to the deanery of Westminster. Under the same patronage Whewell was given the mastership of Trinity College, in preference to Sedgwick who, like Conybeare, was of Whig convictions.

On religion, the central members of the English school were united. They all belonged to the same middle band of opinion within the Anglican Church. Conybeare's son, William John, in his article on 'Church Parties' for the *Edinburgh Review* (1853),

---

[15] 'Additional Notices on the Fossil Genera Ichthyosaurus and Plesiosaurus', *Trans. Geol. Soc.* i (1824), 123.

[16] See Briggs, *The Age of Improvement 1783–1867*, 1959.

[17] See Brock, *The Great Reform Act*, 1973.

divided the Anglican spectrum into eight colours. These he grouped into three categories: Low, Broad, and High Church. The Broad Church was less distinct as a party than the other two, lacking even a periodical. Conybeare the younger characterized the Broad Church men as moderate and tolerant in politics and religion, and supportive of moderate reform.[18]

These characteristics applied to the central members of the English school who represented that liberal Anglican element which pervaded the British Association. Jack Morrell and Arnold Thackray, in their exhaustive study of the first decade of the British Association, describe how it formulated its ideology during the first three meetings at York, Oxford, and Cambridge.[19] It reflected the latitudinarian outlook of Buckland, Conybeare, and especially the men of Trinity College.[20] Inter-denominational co-operation for the advancement of science was encouraged. Quakers and Unitarians were among the most obvious dissenters in the British Association, and they existed also at the fringe of the English school. Conybeare's co-author of the *Outlines of the Geology of England and Wales* was a Quaker. One of the most enthusiastic provincial lecturers on English geology, Robert Bakewell, belonged to the Society of Friends, and other provincial supporters such as Robert Fox or James Prichard had a Quaker background. Davy, to whom the English school was substantially indebted, was of vague denominational allegiance. Even the Roman Catholic cardinal Wiseman used Buckland and de la Beche in his *Twelve Lectures on the Connexion between Science and Revealed Religion* (1836), although with an emphasis on the early, diluvial phase of their geology. The cross-party tolerance was enhanced in the case of Buckland by his irenic emotional constitution; he was on good terms with most people and even friendly with his opponents of the Scottish school.

Buckland and his circle were not opposed, as is commonly believed, by the entire Evangelical Low Church party. Several prominent Evangelicals gave him encouragement, for example Shute Barrington, bishop of Durham and powerful ecclesiastical patron, John Bird Sumner, bishop of Chester and later archbishop of Canterbury, and George Stanley Faber, an early contributor to the *Christian Observer*. Samuel Wilberforce, son of the William who

[18] xcviii (1853), 273–342.
[19] *Gentlemen of Science*, 1981, ch. 5.
[20] Cannon, *Science in Culture*, 1978, ch. 2.

founded the *Christian Observer* and around whose home originated the Evangelical Clapham Sect, was a pupil of Buckland and a life-long ally. Samuel Wilks, a later editor of the *Christian Observer*, used his position to shield the new geology from the criticism of correspondents. Those extremist Evangelicals whom Conybeare the younger called the Recordites did attack Buckland and his circle. Opposition came also from the High Church party, especially from Oxford's Tractarians during the 1830s and early 1840s.

## THE ENGLISH SCHOOL

### Characteristics

Many years ago Archibald Geikie recognized a late eighteenth- and early nineteenth-century Scottish school of geology which he identified with Huttonianism.[21] Roy Porter argues that Hutton was not particularly representative of Scottish geology, and broadens the definition to include the Wernerians as well.[22] The Scottish school was centred on Edinburgh, its Royal Society and its university. John Playfair, professor of natural philosophy, represented Hutton's cause, and Robert Jameson, professor of natural history, adhered for a long time to a traditional form of Wernerian Neptunism. Jameson was also the founder of the Wernerian Natural History Society.[23] David Brewster acted in a variety of editorial positions as a roving ambassador for Scottish geology; for some years he edited, together with Jameson, the *Edinburgh Philosophical Journal*. William Henry Fitton was a pupil of Jameson, and although he joined other Wernerians in the reappraisal of fossils in stratigraphy he remained partial to the Wernerian cause.

Scottish geology was characterized by a marked philosophical stance which contributed to the sterile controversy in which the Huttonians and Wernerians became locked over the origin of basalt and granite. It restricted geology to little more than a study of the mineral kingdom. Moreover, Scottish geology had an overriding commitment to public utility, science being geared to the teaching needs of the university.

[21] *Scottish School of Geology*, 1871. See also his *Founders of Geology*, 1897.

[22] *Making of Geology*, 1977, 149–56. For 'Wernerianism' and 'Huttonianism' see ch. 10 below.

[23] The full extent of Jameson's position, its changes over the years, and his editorial policies still await a major study.

No English school of geology has as yet been formally defined in the secondary literature. English and Scottish geology are often lumped together. However, the notion of British geology does not allow for a historiography which is fine-grained enough to do justice to Buckland and his circle. In this study the existence of an English school and its characteristics are based on the following seven features, discussed and supported by quotations in later chapters. Some features were shared with other schools of geology, but collectively the seven made the English school distinctive.

1. A characteristically English school was recognized by Buckland and his contemporaries. In early correspondence and lecture notes Buckland referred to London and Oxford geology as distinct from the schools of Freiberg and Paris and as presenting a contrast to the Edinburgh School. A sketch of 'the Progress of Geology' from the Greeks to the early nineteenth century ended with the 'comparative merits of the Edinburgh and Oxford School of Geology'.[24] His intention here was in part to distance himself from the conflict between Huttonians and Wernerians; with neither party did Buckland associate, let alone identify. For his lectures he printed a 'Geological Thermometer' which listed the names of all well-known geologists and mineralogists in Europe from the 'hot' extreme of Plutonism to the 'cold' polar opposite of Neptunism. Buckland, Conybeare, and Sedgwick were carefully placed in the very middle, not involved in the excess of Scottish conflict.[25] Conybeare put this formative contrast in the print of his *Outlines*:

The Edinburgh school has to boast of several distinguished geological names; it is impossible to mention that of Playfair without the admiration demanded by a genius of a very high philosophical order, or that of Jameson without the respect due to a long and meritorious career of labours devoted to the advancement of this science; but we cannot but feel the injurious effects which have in this instance been produced by that excessive addiction to the theoretical speculations, which has converted the members of that school into the zealous partisans of rival hypotheses, and led them to contribute far less than they otherwise must have done, to the real progress of inductive geology. To this cause we must ascribe it, that it has fallen so far behind the schools of London and Oxford . . .[26]

[24] Lecture notes, OUM, Bu P.
[25] A slightly different version of the 'Geological Thermometer' was published in the *Edinburgh Phil. Journ.* vii (1822), 376.
[26] xlviii.

During the 1820s the mention of London and Oxford schools was gradually replaced by that of English geology, though the contrast with Edinburgh remained. Conybeare acknowledged and defined an English school in his Report (1832) to the British Association on 'the Progress, Actual State and Ulterior Prospects of Geological Science'. Whewell did the same though more emphatically in the *British Critic*. The Scots too recognized the English school as a concrete and separate entity; Brewster referred to it, for example, in his review of Buckland's Bridgewater Treatise for the *Edinburgh Review* (1837).

2. The English school was characterized by its work on the geology of England and Wales. The older Primary rocks were best known from Germany and from Scotland; the younger Tertiary strata had been described in France, though also in England; but the very thick and complex middle section of the geological record, the Secondary formations, occurred in a uniquely complete and condensed manner within England's borders. The stratigraphy of these rocks was a trophy of the English school.

3. The Secondary strata proved rich in fossils. The discovery of fossil monster reptiles, later known as dinosaurs, was an accomplishment of English geology. Thus the interest and importance of fossils was emphasized, especially for the purpose of stratigraphy. The *Edinburgh Philosophical Journal* for some years did not appreciate the importance of paleontology, and as late as the early 1830s the Scotsman John MacCulloch regarded the proper subject of earth science as mineralogy.

4. The English school was committed to diluvialism during the 1820s and to the progressivist synthesis of earth history in the 1830s. Diluvial geology was a mainly English phenomenon, and progressivism, although shared by many Wernerians, was rejected by the Huttonians. The English school emphasized that progress had been a gradual preparation of the crust and surface of the earth for man's habitation and use. Taxonomic progress, from lower to higher species, was believed to have been a derivative effect of environmental change.

5. The relationship of the new discoveries of geology to biblical history was a matter of deep concern to the English school and was widely discussed. Members of both branches of Scottish geology tended to disapprove of mixing matters of science with those of religious belief; to them biblical exegesis was the territory of theologian, not to be invaded by geologists.

6. English geology aligned itself enthusiastically with the tradition of Paleyan natural theology. The phenomena of geology,
especially those known from paleontology, were presented as new
evidence of design in nature, proofs of God's existence and of his
attributes of power, wisdom, and goodness. Both Buckland and
Whewell wrote Bridgewater treatises. Neither Huttonians nor
Wernerians were explicitly linked to natural theology, although the
argument from design found wide acceptance outside the English
school.

7. The economic aspect of geology was regarded as of little
interest by the English school. They felt it was better left to metropolitan institutions such as the Royal Institution or to provincial
mineral surveyors. Even Buckland, who had a passion for
questions of public utility, repressed this until he began to dis-
associate himself during the 1840s from Oxford and from the circle
of his former colleagues.

A remarkable illustration of the lowly image of economic geology
was Buckland's negative reaction to an invitation to give a lecture
series at the Royal Institution in 1823. The suggestion came from
Buckland's supporters at the Royal Society: Davy, Gilbert, and
Wollaston. He felt obliged to consider the request seriously, and he
wrote to the bishops of London, Durham, and Oxford, to the
chancellor and vice-chancellor of the university, to several heads of
colleges, and to Peel, at the time MP for Oxford, asking if
acceptance might 'compromise the dignity of the University'.[27] The
minutes of the managers' meetings at the Royal Institution record
that Buckland declined the invitation to lecture.[28]

*Representative*

Our view of the early nineteenth century has suffered a major
distortion from the mistaken assumption, that the impact of Lyell's
*Principles of Geology* (1830–33) was such that we must reckon the
years after 1830 'the age of Lyell'.[29] Buckland and his circle valued

[27] E.g. Buckland to Peel, 25 Apr. 1823 and Peel to Buckland, 26 Apr. 1823, BL, Pe P.
[28] *Archives of the Royal Institution of Great Britain*, v (1975), VI-392). See also Berman,
*Social Change and Scientific Organisation*, 1978. Even the Geological Society distanced itself
from its antecedent mineral history interests: see Weindling, 'Geological Controversy
and its Historiography', *Images of the Earth*, ed. Jordanova and Porter, 1979, 248–71.
[29] See n. 7.

Lyell's new observations but did not subscribe to his theoretical speculations. His books were undoubtedly popular: the *Principles* enjoyed a sixth edition only ten years after it was first published, and the *Elements of Geology* (1838) also became a widely used textbook. However, such bibliographical statistics mean little in isolation; they must be compared with the popularity of the literature produced by the English school before any talk of an 'age of Lyell' would have meaning and validity.

Buckland's Bridgewater Treatise, which, as a paleontology textbook, was more specialized than Lyell's, enjoyed nevertheless considerably popularity as well. It did not go through as many editions as Lyell's *Principles*, but a comparison of the number of editions is valueless unless the print-run and the price of each book are identical. The first edition of Lyell's book consisted of only 1,500 copies; the first edition of Buckland's Bridgewater Treatise had an initial print-run of no fewer than 5,000.[30] The advance interest was so great, enhanced by an early review in the *Quarterly Review* (in April 1836, some five months before its appearance), that the entire edition was sold out before any copy had come on the market. A second printing had immediately to be added to meet the demand (with additional notes by Buckland), but even this was soon sold out, and a second edition was published in 1837. The price of Buckland's Bridgewater Treatise was quite high, thirty-five shillings, no cheaper than Lyell's *Principles*.

To this considerable number of copies must be added all the books written by other members of the English school; for example, the *Outlines* by Conybeare and Phillips (1822); de la Beche's *Geological Manual* (1831) and his *Researches in Theoretical Geology* (1834); Phillips's *Guide to Geology* (1834) and his *Treatise on Geology* (1837); even Bakewell's *Introduction to Geology* (1828) and Mantell's *Wonders of Geology* (1838). Some of these texts went through several editions. This selection of general books does not include the more specialized works by Buckland, Murchison, and Sedgwick.

Mantell, a friend of Lyell, put Buckland's Bridgewater Treatise at the top of a list of recommended books in his *Medals of Creation* (1844), and added that copies should be available to every reader 'as they are to be found in every public, and good private library in

---

[30] Wilson, *Charles Lyell*, 270. Buckland to Featherstonhaugh, 18 Apr. 1836, CUL, Se P; Buckland to Irvine, 25 Feb. 1837, CC, Bu P.

the kingdom'.[31] Mantell's pupil Richardson described Buckland's book in his own *Introduction to Geology* (1851) as follows: 'In pictorial beauty, accuracy, and taste, this work is, perhaps, unrivalled in scientific literature.'[32] Buckland himself answered a request for recommended reading by citing Ansted, de la Beche, Mantell, Phillips, and his own Bridgewater Treatise, but added: 'Mr Lyell's works are excellent but too full of shifting hypotheses and bold theory for a beginner.'[33]

At the popular level the geology of the English school was expounded in a variety of periodicals. At a still lower level clergymen and itinerant lecturers spead the new earth history to towns and villages where local papers printed summaries of their presentations. Innumerable instances could be cited; but the following two examples will suffice. In 1837 a Revd Hinton delivered three lectures on geology in the town hall at Wallingford; he discussed such topics as the order of succession of rock formations and fossils, volcanic phenomena, and especially Buckland's hyena den theory. In 1837 a Dr Lloyd gave a series of six lectures at the Mechanics' Institute in Leamington, extensively reported in the *Leamington Chronicle*, in which Buckland's Bridgewater Treatise was discussed at length.

George Eliot's *The Mill on the Floss* (1860) illustrates how Buckland's book may have come to people's attention in many an English village. Eliot, who herself had read the Bridgewater Treatise, mentioned it in connection with a ladies' book club:

Did Lucy intend to be present at the meeting of the Book Club next week? was the next question. Then followed the recommendation to choose Southey's Life of Cowper, unless she were inclined to be philosophical and startle the ladies of St Ogg's by voting for one of the Bridgewater Treatises. Of course Lucy wished to know what these alarmingly learned books were, and as it is always pleasant to improve the minds of ladies by talking to them at ease on subjects of which they know nothing, Stephen became quite brilliant in an account of Buckland's Treatise, which he had just been reading.[34]

[31] i. 11.
[32] 69.
[33] 14 Dec., year omitted, OUM, Bu P.
[34] Vol. iii, bk. 6, ch. 2. See also Eliot to Jackson, 4 Mar. 1841, *George Eliot Letters*, ed. Haight, viii (1978), 8.

OXBRIDGE REFORM

*Edinburgh criticism*

Most historians dealing with the geology of the early nineteenth century have interpreted the work of Buckland and his circle in the light of pan-European intellectual debates engendered by the secular philosophy of the Enlightenment. In this international context Buckland's geology has been disparagingly characterized as a form of antiquated science in the service of supposed religious truth.[35]

The English school, however, can only be adequately and accurately explained in the limited and specific setting of England itself, and more particularly of its ancient universities. Here Buckland and Sedgwick attempted to acquire an audience for the new geology. Interest in the relationship of geology to questions of natural and revealed religion did not spring from religious ortho-doxy, but arose from the need to gear the teaching of the new science to the requirements of Oxford and Cambridge as centres for the education of Anglican clergy. The value of geology was judged by its congruence with the existing tradition of learning and by its relevance to a clerical education. This institutional constraint was a more immediate determinant of geological theory than any pan-European controversies.

The interest in geology at England's ancient universities in the early nineteenth century was bound up with the internal reform of the curriculum. From about 1800 Oxford and Cambridge began to awake from a slumber of intellectual indolence. The standard of traditional education in classics and mathematics was raised and a curiosity was aroused for new areas of study, for example geology, political economy, and Sanskrit. The remarkable popularity of the lectures by Buckland and Sedgwick was a symptom of this new spirit.

Most college fellows were by profession clergymen, not teaching dons. Their fellowships prevented them from getting married, and the majority of fellows hoped and waited for some form of advance-ment to ecclesiastical preferment, of which the colleges controlled a vast network. Desirable positions were, apart from the headship of

---

[35] E.g. Gillispie, *Genesis and Geology*, 1951; id., *Edge of Objectivity*, 1960. Glass *et al.* (eds.), *Forerunners of Darwin*, 1959.

colleges, a college living, a canonry, a deanery, or a bishopric. Buckland, Conybeare, Sedgwick, and Whewell all depended for the main part of their income on one or more of these ecclesiastically controlled positions. Some professorial chairs also were rich prizes, but not all. Buckland's annual stipend for each readership was no more than £100. Sedgwick's professorial salary was increased to a mere £200.

This contrasted with the organization of the Scottish universities. Among these Edinburgh attained a pre-eminence in science toward the end of the eighteenth century. This university was not primarily an Anglican seminary like Oxbridge, but a non-denominational and non-collegiate institution which emphasized professional teaching by professorial class-room instruction, much after the manner of continental universities. Classics and mathematics were cultivated, but Edinburgh also had a medical school of international repute, established by pupils of Boerhave from Leiden. As part of the medical training such sciences as anatomy, botany, and chemistry flourished. Thus science existed in its own right and was not tailored to the requirement of a clerical education.

John Merz and many historians since have attributed the eminence of Edinburgh science to this difference in organization between English and Scottish universities.[36] In Scotland the famous names of science tended to be university professors; in England much of the best chemistry and physics was done outside the universities. Jack Morrell attributes the excellence of Edinburgh science to the professorial system, which was supported by a class fee system.[37] The basic annual salary of professors was low, and they had to supplement this with student fees. This functioned as a form of payment by result and popularity: the novelty, quality, and professional utility of a lecture series determined the size of the audience and thus the professor's income. Various Scottish writers on geology of the early nineteenth century held such positions, e.g. John Playfair, Robert Jameson, and later also John Fleming.

Scottish pride in the pre-eminence of Edinburgh learning combined with an envy of Oxbridge privilege was expressed in the *Edinburgh Review* (begun in 1802). Although intended as a non-partisan organ it rapidly became a vehicle of Whig reform causes. During the early part of the nineteenth century the *Edinburgh Review*

---

[36] *History of European Thought in the Nineteenth Century*, i (1896), ch. 3.

[37] 'The University of Edinburgh in the Late Eighteenth Century', *Isis*, lxii (1971), 158–71.

launched a sustained attack on Oxbridge, though more particularly on the older of the two, Oxford. The criticism focused on the differences between the Scottish and English universities. The classical system was attacked as backward, the lack of excellence in science was emphasized, tutorial teaching was criticized and the professorial system advocated, and the Anglican exclusiveness of Oxbridge was denounced. The first salvos were fired in the period 1808 to 1810, mainly by John Playfair.[38] Sniping continued during the 1810s and 1820s. In 1821 Daniel Keyle Sanford, professor of Greek in Glasgow, sneered that Oxford was just beginning to recognize that human knowledge might go further than the peaks which Bently or Scaliger had scaled.[39]

Criticism reached a new height in the early 1830s when William Hamilton, a Glaswegian who had been refused a fellowship at the end of his studies at Balliol College, joined the Edinburgh reform movement. In 1831 he wrote for the *Edinburgh Review* 'On the State of the English Universities, with more especial reference to Oxford', charging that Oxford had fewer teachers with a wide reputation than any other university; the pre-eminence of the tutorial over the professorial system he regarded as 'vicious'. Hamilton added a further attack on the Anglican exclusiveness of Oxbridge in a review 'On the Right of Dissenters to Admission into the English Universities' (1834).[40]

Competition to Oxbridge drew closer with the establishment of the University of London (1826; later University College) organized on continental and Scottish lines. Moreover, Charles Babbage, himself a Cambridge man, published a controversial jeremiad on a supposed *Decline of the State of Science in England* (1830), echoed the following year by Brewster in his *Edinburgh Journal of Science* and aimed at Oxbridge, the Royal Society, and Peel's Tories.[41]

## Geology as apologia

Reaction ranged from the founding of the *Quarterly Review* (begun in 1809) to that of King's College (1829) in defence of Anglican and

[38] xi (1808), 249–84; xiv (1809), 429–41; xvi (1810), 158–87; xvii (1810), 122–35; xxvii (1816), 87–98.
[39] xxxv (1821), 304.
[40] liii (1831), 384–427; liv (1831), 478–504; lix (1834), 196–227; lx (1834), 202–30; lx (1835), 422–45.
[41] v (1831), 1–16.

Tory values. The criticism of northern Whigs may also have hastened the pace of internal reform at Oxbridge.[42] However, before the attack in 1808 some reform had already started. The mathematical tripos at Cambridge founded in the middle of the eighteenth century initiated competitive examinations. Half a century later Oxford followed suit; after its change in the examination statutes and the introduction of an honours school in 1801 (revised in 1807) study and teaching became an increasingly serious affair. Nevertheless, for a long time science remained entirely optional and subsidiary to the *literae humaniores,* i.e. classics with some history, philosophy, languages, and literature.[43]

One Oxonian who warmly supported the new examination statutes was Edward Copleston, provost of Oriel College from 1814 to 1828, when he left to become bishop of Llandaff. But Copleston vigorously defended the Oxford system of classical education in three replies to the *Calumnies of the Edinburgh Review against Oxford* (1810, 1811). Although he was committed to the improvement of Oxford education he rejected the idea of a radical overhaul. Grenville, to whose election as chancellor of the university in 1809 Copleston had contributed, complimented him for his valiant defence of Oxford's tradition of education and learning. Neither man, however, was for the status quo; both believed in the introduction of new subjects, though in a complementary role to classical studies. Copleston extended the practice of competitive examinations when as provost of Oriel he instituted a system of fellowships by examination. As a result he gathered around him a group of young men interested in academic excellence. The roll included such names as Arnold, Hampden, Keble, Newman, Pusey, Whately, and Wilberforce. Copleston, together with most of his talented and aspiring men, attended Buckland's geology lectures.

The fortunes of geology at Oxford and Cambridge were intimately interwoven with the issues of reform. The introduction of geology was carefully tailored to fit the traditional content and purpose of a university education. This was the main motivation for Buckland and Conybeare to formulate an Oxford school of geology in contrast to Edinburgh geology. Among the characteristic

[42] Engel, 'From Clergyman to Don', Ph.D. thesis, 1975 (OUP book, forthcoming). See also his 'Emerging Concept of the Academic Profession at Oxford', in Stone (ed.), *University in Society,* i (1975), 305–52. Garland, *Cambridge before Darwin,* 1980.

[43] Ward, *Victorian Oxford,* 1965. Winstanley, *Early Victorian Cambridge,* 1940.

features of English geology were those which made the subject congruent with and complementary to the tradition of classical learning and which justified its inclusion in the curriculum for the education of the Anglican clergy. Thus the evidences of the deluge were emphasized in the 1820s, and the relationship of geology and the creation story in the 1830s. Moreover, science was presented as a subject which demonstrated divine design in nature, as natural theology, needed as part of a proper clerical education. Both Buckland and Sedgwick were rewarded with ecclesiastical preferment; Buckland was made a canon of Christ Church in 1825, which gave him agreeable living-quarters and over £1,000 a year; Sedgwick's appointment as prebendary of Norwich in 1834 gave him financial independence also.

The Oxbridge devotees of science took part in the defence of the classical system against the *Edinburgh Review*, even though some of them held Whig convictions. Kidd wrote an *Answer to a Charge Against the English Universities* (1818). Buckland emphasized that geology should merely be grafted 'on that ancient and venerable stock of classical literature from which the English system of education has imparted to its followers a refinement of taste peculiarly their own'.[44] Conybeare believed that geology and a classical education could be combined to mutual advantage.[45] Lyell concluded in a review of the 'State of the Universities' for the *Quarterly Review* (1827) 'that no extensive or violent changes are required, in order to accommodate, in a very short time, the institutions of Oxford and Cambridge to the wants and spirit of the present age'. In support of this Lyell cited several lecture courses. 'The most popular courses, however, have been those of the professors of geology, Dr. Buckland at Oxford and Mr. Sedgwick at Cambridge; they have continually attracted as many as the moderate dimensions of their class-rooms could contain.'[46] Whewell also held up Oxbridge geology as a defence against Edinburgh criticism. In a review of 'Science of the English Universities' for the *British Critic* (1831), he argued that the charge of the neglect of science was false and that the reverse was true, especially in view of the geology of Buckland, Conybeare, and Sedgwick:

[44] *Vindiciae Geologicae*, 2–3.
[45] Conybeare and Phillips, *Outlines*, xlviii.
[46] xxxvi (1827), 257.

Indeed, we are compelled to declare, that we know not in what part of England, *except in the two Universities*, a young student has any possibility of acquiring a knowledge of Geology according to the views of those who have successfully cultivated it in modern times. And we do think it strange, that the Universities, which thus unquestionably have been and are the first and most zealous nurses of this youngest of the sciences, should be exposed to the trite and unmeaning charge of cherishing antiquated dogmas and neglecting the advances of modern discovery.[47]

The same apologia was given by Oxford's conservative *University Magazine* (1834).

At Edinburgh too the geology professors were allied to the indigenous ideology of university education and reform. Playfair, who initiated the *Edinburgh Review* attack on Oxbridge, was the man who had popularized uniformitarianism in *Illustrations of the Huttonian Theory of the Earth* (1802). The *Edinburgh Review* criticized the connection of geology with issues of biblical history and exegesis in Buckland's *Reliquiae Diluvianae* (reviewed by Fitton), and in his Bridgewater Treatise (by Brewster). Conversely, this very aspect was appreciated in a review of the *Reliquiae* for the *Quarterly Review* (by Copleston). The Scottish criticism came from Huttonians as well as Wernerians. Fleming attacked Buckland's geology in a series of papers carried by the *Edinburgh Philosophical Journal*, edited by Brewster and Jameson. The latter's Wernerian Natural History Society was used for the same purpose. The emphasis on a characteristically English school of geology was to a large extent determined by its allegiance to Oxford and Cambridge in its defence against the attacks in the Edinburgh journals.

In conclusion, the context of pan-European geological controversies does not provide a proper basis for an understanding of the work of Buckland and his circle. Such understanding will come only when one recognizes the existence of an Oxford and English school of geology which formed in contrast to Scottish geology and when both are seen as allied to the indigenous ideology of university education exemplified by respectively Oxbridge and Edinburgh.

## CHANGES

The Oxford and English school experienced its heyday in the 1820s and 1830s. During these decades its characteristics were by no

[47] ix (1831), 74.

means static. Considerable changes occurred over the years. Accordingly, this study has been divided into three parts, each dealing with a different aspect of English geology which was the focus of interest at a different time. The first part discusses diluvialism, the second the progressivist synthesis, and the third the relationship of geology with natural theology.

In the 1840s the fabric of the English school disintegrated and it lost several of its characteristic features. The contrast with Scottish geology dissolved; the issue of organic evolution led to a realignment which brought members of the English school and Scottish Evangelicals together. Other interests replaced the old ones: de la Beche and Buckland became increasingly committed to economic geology. Murchison and Sedgwick developed a bitter controversy over questions of stratigraphic subdivision. Whewell and Conybeare effectively abandoned their former geological hobby. Buckland's old allies opposed him when he attempted to introduce Agassiz's glacial theory.

# PART I

# HYENA DENS AND THE DELUGE

## *Diluvial Geology as an Adjustment to Oxford Learning*

We, we shall view the deep's salt sources poured
Until one element shall do the work
   Of all in chaos; until they,
   The creatures proud of their poor clay,
Shall perish, and their bleached bones shall lurk
   In caves, in dens, in clefts of mountains, where
The deep shall follow to their latest lair;
   Where even brutes, in their despair,
Shall cease to prey on man and on each other . . .
Byron, Heaven and Earth,<br>I iii. 170–8

## 2
# Bone Caves and Diluvial Phenomena

Buckland had been interested in cave paleontology long before the early 1820s. During his continental journey of 1816 he had explored some of the famous caves in Germany. The story of Kirkdale Cave, however, began to unfold in 1821. In July of that year quarrymen came across the opening of a cave in the oolitic limestone of Kirkdale, a village near Kirby Moorside in Yorkshire. The cave contained a large quantity of fossil teeth and bones and, believing that these were the remains of cattle, the quarrymen discarded the fossils and scattered them across the roads. Here they were noticed by a physician from Kirby Moorside who, along with several others, collected part of the rich assemblage and distributed specimens among metropolitan scientific institutions, such as the Royal College of Surgeons.

Buckland was first told about the discovery of Kirkdale Cave by Edward Legge, the bishop of Oxford and warden of All Souls, who used to attend the geology lectures. He passed the news on to Joseph Pentland who in turn urged Buckland to visit Kirkdale Cave and collect as many fossils as possible to assist Cuvier in the execution of the second edition of his monumental *Recherches sur les ossemens fossiles* (1821–4): 'I this morning received your kind letter of the 18th inst., and immediately communicated its contents to Mr Cuvier, who desires me to write to you in all haste in order to request you to procure for him if possible some of the fossil bones lately discovered in such abundance in Yorkshire, especially those of the Hyena, as he is now engaged in that part of his new work which treats of fossil Carnivores.'[1]

In December 1821 Buckland went to the site, examined the cave, collected specimens, and in February 1822 announced a theory which thrilled London and Paris alike: not only had Kirkdale Cave been the den of antediluvian hyenas, but the dietary habits of these

[1] Pentland to Buckland, 26 Nov. 1821, Sarjeant and Delair, 'An Irish Naturalist in Cuvier's Laboratory', *Bull. Brit. Mus. (Nat. Hist.)*, Hist. Ser. vi (1980), 284.

animals had produced the seemingly haphazard assemblage of teeth and bones.

Kirkdale Cave was relatively small, some 300 feet long and from two to five feet in width and height. The fossils were embedded in its floor, preserved in a layer of mud which had become partially covered by stalagmite. Buckland determined that the teeth and bones belonged to a non-symbiotic mixture of both carnivorous and herbivorous animals, namely of hyena, elephant, rhinoceros, hippopotamus, horse, ox, two or three species of deer, bear, fox, water rat, and various birds. The first four represented species which are now extinct and genera which at present inhabit tropical regions.

The mere description of the cave and its vertebrate fossils put Buckland on a par with continental comparative anatomists. But he went further and developed an ecological theory to account for the diversity of fossil species inside the narrow cave. The notion that caves were the dens of such animals as bears or hyenas was not novel. Buckland must have been familiar with it from German cave paleontology.[2] However, a mixture of fossils such as that in Kirkdale Cave at first seemed to suggest that the animals had either sought refuge in the cave from a cataclysmal inundation or that their carcasses were swept into it by the swirling waters. The occurrence of fossil bones in caves elsewhere had been interpreted as evidence of a deluge.[3] On 18 November 1821, the day Buckland received a collection of fossils from Kirkdale Cave, he wrote to Pentland that 'it is not easy to conceive that anything short of the common calamity of a simultaneous destruction could have brought together in so small a compass so heterogeneous an assemblage of animals as we here find intombed in a common charnel house, animals which no habit or instinct we are acquainted with could ever have associated with a den of hyaenas.'[4]

However, even before Buckland had visited Kirkdale Cave, he began to change his mind and to discount a cataclysmal cause. On

---

[2] The idea had been expressed by Johann Christian Rosenmüller, *Beiträge zur Geschichte und nähern Kenntniss fossiler Knochen*, 1795. See also Heller, 'Die Forschungen in der Zoolithenhöhle bei Burggaillenreuth von Esper bis zur Gegenwart', *Erlanger Forschungen*, v (1972), 7–56.

[3] E.g. by Johann Friedrich Esper and Georg August Goldfuss. See the 'Einführung' by Geus to the 1978 facsimile of Esper's *Ausführliche Nachricht von neuentdeckten Zoolithen unbekannter vierfüssiger Tiere*, 1774.

[4] Cu P, MS 627, fo. 22.

26 November 1821 he wrote to Miss Talbot about the fossils other than those of hyenas:

How the latter got there is not easy to be conceived unless they be either the wreck of the hyenas' larder or were drifted into a fissure by the diluvian waters. Both are possible cases, but the latter assumes that there was a fissure open at the top, and the account as I at present have of it states that the aperture is a cavern covered all over at the top with continuous beds of limestone and, if so, we can only suppose the bones to be the wreck of animals that were dragged in for food by the hyenas. Not that I suppose a hyena could kill an elephant etc. but that, as we know they do not dislike putrid flesh, we may conceive they took home to their den fragments of these larger animals that died in the course of nature, and which from their abundance in the deluge gravel we know to have been the antediluvian inhabitants of this country.[5]

His belated visit in December 1821 to Kirkdale Cave put into Buckland's hands a number of facts which supported the ecological theory outlined in his letter to Miss Talbot. Buckland observed that the fossil bones had, for the most part, been broken and gnawed; some broken bones exhibited teeth-marks made by hyenas, a fact he illustrated by experiments with live hyenas (figure 5). When a Cape hyena in a travelling collection happened to pass through Oxford Buckland gave it the shin bone of an ox and observed the beast break and gnaw it. 'I preserve all the fragments and the gnawed portions of this bone for the sake of comparison by the side of those I have from the antediluvian den in Yorkshire: there is absolutely no difference between them, except in point of age.'[6]

Further, Buckland identified a substance which he believed to be faecal matter of hyenas, called *album graecum* because of its white colour, indicating that the hyenas had indeed eaten bones inside Kirkdale Cave. This was confirmed by William Wollaston who made a chemical analysis of the substance, and by the keeper of the menagerie at Exeter Change who recognized it as being similar to the faeces of the Cape hyena in his collection. But Wollaston cautioned Buckland by writing that 'though such matters may be instructive and therefore to a certain degree interesting, it may as as well for you and me not to have the reputation of too frequently and too minutely examining faecal products'.[7]

[5] NMW, Bu P.
[6] *Reliquiae Diluvianae*, 1823, 38.
[7] Wollaston to Buckland, 24 June 1822, RS, Bu P.

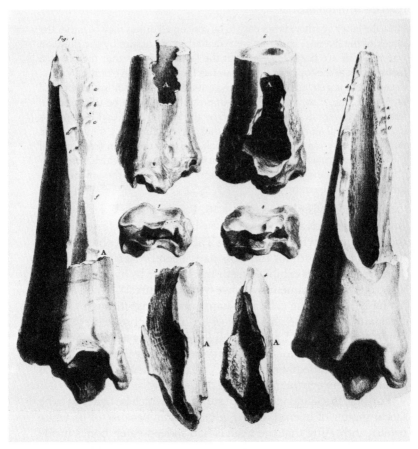

5. Matching teeth-marks (A, a, b, c, d, e, f) made by hyenas on recent ox bones (1, 3, 5) and on fossil bones from Kirkdale Cave (2, 4, 6); recent (7) and fossil scaphoid bone (8) both left untouched.

Another fact which Buckland used in support of his theory was that a number of bones appeared polished and worn on one side only. This he attributed to the friction generated when the hyenas trod and rubbed their skin over the bones stuck on the floor of the den. With a characteristic ability to cite striking parallels Buckland wrote: 'The same thing may be seen on marble steps and altars, and even metallic statues in places of worship that are favourite objects of pilgrimage: they are often deeply worn and polished by the knees, and even lips of pilgrims, to a degree that, without experience of the fact, we could scarcely have anticipated.'[8]

It seemed, therefore, that the assemblage of fossils had not been produced instantaneously by a cataclysmal event, but was the result of a gradual accumulation over a long period of time, during which hyenas had inhabited Yorkshire and dragged their food into Kirkdale Cave.

In the first public report on the cave which Buckland presented to the Royal Society (7, 14, and 21 February 1822) he discussed a number of other caves including examples from near Plymouth.[9] The quarrying operations carried on in the limestone of Oreston to supply material for the Plymouth breakwater disclosed, from 1816 to 1822, a number of bone caves. Studies of these were communicated to the Royal Society by Everard Home,[10] and possibly because of this Buckland also presented the results of his cave studies to the Royal Society rather than to the Geological Society. He wrote to Greenough: 'The Kirkdale story is too long for any letter and is expanded into a long paper which I shall send directly to the Royal Society to set right all errors that have appeared in their volumes from Plymouth and elsewhere about elephants, rhinoceros, and animals of that kind which are part of the same story as that in Yorkshire and which I flatter myself I have completely made out.'[11]

Approval of Buckland's hyena den theory quickly followed, judging by the number of science journals which reported it, both at home and abroad. Apart from its publication in the *Philosophical*

---

[8] *Reliquiae Diluvianae,* 32.

[9] 'Account of an Assemblage of Fossil Teeth and Bones of Elephant, Rhinoceros, Hippopotamus, Bear, Tiger, and Hyaena, and Sixteen other Animals; Discovered in a Cave at Kirkdale, Yorkshire, in the Year 1821', *Abstract of the Papers, Phil. Trans.* ii (1823), 165–7; *Phil. Trans.,* cxii (1822), 171–236.

[10] E.g. 'An Account of Some Fossil Remains of the Rhinoceros', read 27 Feb. 1817, *Abstract of the Papers, Phil. Trans.* ii (1823), 66–7; ibid., cvii (1817), 176–82.

[11] 13 Jan. 1822, CUL, Gr P.

*Transactions,* the *Annals of Philosophy* (1822) and the *Philosophical Magazine* (1823) reported Buckland's theory in England; in Scotland the *Edinburgh Philosophical Journal* (1822, 1823) felt compelled to do the same; in Germany it was published in the *Notizen aus dem Gebiete der Natur- und Heilkunde* (1822), and in France in both the *Annales de chimie et de physique* (1823) and the *Journal de physique, de chimie, d'histoire naturell et des arts* (1823).

The novelty of Buckland's 'hyena story' caused some amused disbelief or even ridicule.[12] But in the fourth volume of the *Recherches sur les ossemens fossiles* (1823) Cuvier sanctioned the theory and commented on Kirkdale Cave: 'Visitée aussitôt après sa découverte par plusieurs hommes instruits et surtout par le savant et ingénieux géologiste M. Buckland, on n'a rien à désirer à son sujet.'[13] Cuvier expressed his complete accord with Buckland's theory, changing his mind on a number of disagreements he had initially aired. The Royal Society awarded Buckland its highest honour in the form of the Copley Medal (1822), never before given to geology. Its president, Humphry Davy, complimented Buckland on his report: 'I do not recollect a paper read at the Royal Society which has created so much interest as yours.'[14] In his presentation speech Davy accurately assessed the significance of Buckland's cave work as a major contribution to historical geology; it was the first study of a separate period in the history of the earth: 'I cannot conclude this part of my subject without congratulating the Society, that by these inquiries, a distinct epoch has, as it were, been established in the history of the revolutions of our globe: a point fixed, from which our researches may be pursued through the immensity of ages, and the records of animated nature, as it were, carried back to the time of the creation.'[15]

Buckland continued to explore other bone caves. At home he examined such new examples as Goat Hole at Paviland in which a human skeleton was discovered. He made another journey to Germany in 1822 to study a number of its caves. His cross-section of the cave at Gaylenreuth in Bavaria became the icon of cave paleontology. This and other German caves, however, had not been dens

---

[12] *Gentlemen's Mag.* xcii (1822), 491–4.

[13] 'Immediately visited after its discovery by several competent men and above all by the learned and skilled Mr Buckland, there is nothing left to be desired on this subject.' 302.

[14] 18 Mar. 1822, RS, Bu P.

[15] Davy, *Six Discourses,* 1827, 51.

of hyenas but of bears. Buckland estimated that bears had occupied the cave at Kühloch in Bavaria for a period of some 1,000 years. From the earlier literature Buckland collected data about caves, going back to Johann Friedrich Esper, Georg August Goldfuss, Johann Christian Rosenmüller, and even to Leibniz.

In addition to caves Buckland recognized the existence of potholes or fissure caves open to the surface. Bone breccias in such fissures were known from various parts of Europe, in particular from Gibraltar. Buckland attributed the origin of such accumulations to the possibility that animals had fallen in by accident. He cited a contemporary analogue, a chasm at Duncombe Park near Kirby Moorside where in recent times a mixture of such species as dog, goat, sheep, and hog had accumulated, after falling to their deaths. Such bone breccias lacked the evidence of habitation by hyenas, such as gnawing-marks.

This new work, combined with the earlier report to the Royal Society, was turned into the first part of Buckland's classic *Reliquiae Diluvianae* (1823). The first edition of 1,000 copies sold out very rapidly, and so did the second edition (1824), another 1,000. The book received lengthy reviews in both the *Quarterly Review* (by Copleston) and the *Edinburgh Review* (by Fitton). Across the Atlantic, Silliman's *Journal* devoted two lengthy review instalments to the *Reliquiae Diluvianae*. The *Annals of Philosophy* carried extracts from it. A variety of lesser periodicals marvelled at it.

The discovery of Kirkdale Cave had come as an interruption of Buckland's work on traditional diluvial phenomena. These phenomena had been discussed by a succession of authors from Alexander Catcott on, and at Oxford they had been competently described by Kidd in his *Geological Essay* (1815). In an appendix to his inaugural lecture (1819) Buckland had presented a summary of diluvial evidence. Broadly speaking, these consisted of (1) the unsorted deposits of clay, gravel, and, in places, huge boulders strewn across hills and valleys, and (2) the shape of hills and valleys.

Buckland devoted a considerable amount of field work to the question of the origin of gravel deposits and of valleys, the result of which he communicated to the Geological Society. His 'Description of the Quartz Rock of the Licky Hill in Worcestershire' (1819) identified this hill as the nearest possible source of a large quantity of siliceous pebbles which occur as far south-east as London. By tracing their dispersal Buckland believed he mapped the path of a diluvial current which had excavated valleys in its way.[16] Later papers on the formation of valleys in the south of England (1824 and 1825) recognized in addition to a presumed diluvial tide such influences as elevation, faulting, and fluvial erosion, contributary to shaping valley morphology.[17]

Part of this work was included in the second part of the *Reliquiae Diluvianae*. Buckland defined the diluvial gravel (Diluvium) and distinguished it from deposits formed since the flood (Alluvium). This distinction, made in colaboration with his former student Strangways, who worked in Russia near St. Petersburg, gave stratigraphic identity to the diluvial gravel and its fossil fauna, as the record of the antediluvial period of earth history:

As I shall have frequent occasion to make use of the word *diluvium*, it may be necessary to premise, that I apply it to those extensive and general deposits of superficial loam and gravel, which appear to have been produced by the last great convulsion that has affected our planet; and that with regard to the indications afforded by geology of such a convulsion, I entirely coincide with the views of M. Cuvier, in considering them as bearing undeniable evidence of a recent and transient inundation. On these grounds I have felt myself fully justified in applying the epithet *diluvial*, to the results of this great convulsion; of *antediluvial*, to the state of things immediately preceding it; and *postdiluvial*, or *alluvial*, to that which succeeded it, and has continued to the present time.[18]

Buckland's familiarity with the vertebrate fossils of the gravel had facilitated the identification of the cave fossils. In his *Reliquiae Diluvianae* he pointed out that the fossil species from the gravel and from the cave mud were identical and had lived during the same period of antediluvial time. He speculated that the diluvial current had simultaneously emplaced the gravel deposits and caused the extinction of the cave fauna, also depositing the layer of mud in which its fossils had been preserved. One cave fossil which initially had not been found in gravel deposits was the hyena. But Buckland predicted that it would be discovered, and sure enough the jaw and other parts of a hyena were in due course found near Rugby; later the skull also turned up. This discovery

[16] *Trans. Geol. Soc.* (Ser. 1), v (1821), 506–44.
[17] 'On the Excavation of Valleys by Diluvial Action', ibid. (Ser. 2), i (1824), 95–102; 'On the Formation of the Valley of Kingsclere', ibid., ii (1829), 119–30.
[18] *Reliquiae Diluvianae*, 2.

was the occasion of apocalyptic verse on 'The Last British Hyaena' (quoted in ch. 5).

It must be obvious, though, that the implications of Buckland's hyena den theory were at variance with the diluvial theory and with the title of his book, which misled those who were not familiar with its conclusions. The cave fossils were not the 'relics of the deluge', but those of the feeding habits of hyenas accumulated over a long period of antediluvial time. The title was not so much a summary or conclusion of the book's contents as their disguise or apology.

The hyena den story undermined traditional diluvialism in several ways. In the first place it detracted from eighteenth-century Mosaical geology which ascribed the deposition of most fossiliferous and stratified sediments to the waters of the deluge. This view survived in England into the early nineteenth century and was expressed in the first volume of Parkinson's influential *Organic Remains of a Former World* (1804). However, by the time Parkinson published the third volume of his *magnum opus* in 1811, he had changed his mind and adopted Cuvier's view of earth history.

The English understanding of Cuvierian geology was based on Kerr's translation (1813) of the preliminary discourse to the *Ossemens fossiles* (1811), edited by Jameson. It visualized not only two periods of earth history, ante- and post-diluvial, but several more prior to the creation of man. Each of these was characterized by different faunal communities or 'worlds'. Most fossiliferous rocks were deposited gradually during these epochs. They were terminated by geological upheavals which caused the extinction of many forms of life. The last of these upheavals was identified by Jameson as the Mosaical deluge. Buckland's hyena den theory gave vivid reality to the antediluvial period; it also limited the effect of the biblical deluge to the emplacement of loose surface sediment.

Secondly, Buckland's cave work reduced the power of the deluge to transport these surface deposits. The presence of tropical fossils in the gravel was traditionally seen as proof of the power and universality of the deluge; the animals had been swept away from southern regions to the north over a distance of thousands of miles. Even before the discovery of Kirkdale Cave, Buckland doubted whether the gravel fossils originated in equatorial latitudes. Pentland worried that such a view would appear contrary to the notion of a universal deluge: 'if you suppose that the bones of those animals have been deposited where we find them or nearly so, by their former possessors, it will be a strong argument in favour of those who suppose that the last deluge (that consequently which is recorded in the Mosaic History), was rather partial or restrained to certain countries, than general over the entire earth's surface'.[19] The hyena den theory suggested that the tropical animals had indeed lived and roamed in Yorkshire and consequently in northern latitudes.

Thirdly, and most importantly, Buckland's cave work undermined the widely held theory of how the deluge had happened, namely by an interchange of dry land and the sea. This view was championed by Deluc who counted among his many supporters Henry de la Fite from Trinity College in Oxford, who translated Deluc's *Elementary Treatise on Geology* (1809). A passage in it described how 'this dreadful catastrophe had been caused by the sinking of an immense extent of land, whence a *new land* was formed from what before was the *bed* of the *sea* . . .'[20] This theory of the deluge had the sanction of Cuvier. However, the hyena den theory implied that antediluvial animals had lived on land which was dry both before and after the flood. An interchange of land and sea could not, therefore, have occurred. Buckland pointed out that an:

important consequence arising directly from the inhabited caves, and ossiferous fissures, the existence of which has been now shown to extend generally over Europe, is, that the present sea and land have not changed place; but that the antediluvian surface of at least a large portion of the northern hemisphere was the same with the present; since those tracts of dry land in which we find the ossiferous caves and fissures must have been dry land also, when the land animals inhabited or fell into them, in the period immediately preceding the inundation by which they were extirpated.'[21]

Buckland did not speculate about the cause of the deluge, but he apparently visualized it as a giant surge or tidal wave.

The notion that a diluvial current had been responsible for the dispersal of erratic boulders had been given respectability by the Genevan naturalist Horace-Bénédict de Saussure in his *Voyages dans les Alpes* (1780–96). He described the occurrence of gravel and boulders in the Alps, in particular in the vicinity of the Lake of Geneva, and concluded 'que c'est une grande débâcle, ou un

[19] Pentland to Buckland, 6 Nov. 1820, Sarjeant and Delair, 'Irish Naturalist', 264.
[20] 390–1.
[21] *Reliquiae Diluvianae*, 162.

courant d'une violence et d'une étendue considérable, qui les a transportés et déposés dans leurs places actuelles.'[22] When John Ruskin visited the area some decades later the erratics inspired him to verse which portrayed the aquatic débâcle as diluvial 'monster surges'.[23] This mechanism was suggested by James Hall in a paper read to the Royal Society of Edinburgh (1812) in which he speculated that the sudden emergence of part of the sea floor had produced tidal waves.[24] Another mechanism to generate giant surges or even initiate an inversion of land and sea was the near approach of a comet. Greenough appeared to give tacit approval to this cometary mechanism; others such as Henslow discussed it,[25] but most members of the Geological Society were disinclined to deal with anything so speculative as the cause of the deluge.

In one respect Buckland's hyena den theory did not detract from traditional diluvialism, but rather added to the geological significance of the diluvial event. If tropical animals had lived in northern latitudes, 'a probable change of climate in the northern hemisphere'[26] seemed to follow.

---

[22]'that a great débâcle or a current of enormous power and extent has transported and emplaced these'. i. 220.

[23] *Works of Ruskin*, ii, *Poems*, ed. Cook and Wedderburn, 406–7.

[24]'On the Revolutions of the Earth's Surface', *Trans. Roy. Soc. Edinburgh* vii (1815), 139–212.

[25]'On the Deluge', *Ann. Phil.* vi (1823), 344–8.

[26]*Reliquiae Diluvianae*, 162. For secondary literature on Buckland's cave studies see Boylan, 'William Buckland, Pioneer in Cave Science', *Studies in Speleology* i (1967), 237–53; and Orange, 'Hyaenas in Yorkshire', *History Today* xxii (1972), 777–85.

# 3
# Objections by Biblical Literalists

To appreciate the quality of Buckland's work on bone caves we must not compare it with geological work done later in the century, not even with his own, but put it in the context of the literature on the subject of the 1820s. By later standards Buckland's *Reliquiae Diluvianae* may have appeared traditional, but during the decade of its appearance it represented modern geology. Some historians have argued that Buckland perpetuated the conservative part of scientific and theological opinion.[1] But the very opposite is true. Fundamentalism, in the sense of an adherence to the literal meaning of the book of Genesis and its geological implications (e.g. the earth only a few thousand years old, much of the crust of the earth deposited during the deluge), began to form a coherent body of literature, a connected bibliography, specifically in opposition to Buckland's *Reliquiae Diluvianae*.

The publication of this work was followed by a stream of articles and books in which Buckland's hyena den theory and his diluvialism were fiercely attacked and in which various alternative theories were proposed, each in accord with fundamentalist belief, or what may be more properly called 'biblical literalism'.[2] Later in the century fundamentalist works such as Philip Henry Gosse's *Omphalos* (1857) were part of the connected bibliography begun by the anti-Buckland literature. The wave of criticism persisted through the 1820s and early 1830s, only to be subsumed in a larger wave which followed the publication of Buckland's Bridgewater Treatise (1836) and, to a lesser extent, the *Discourse on the Studies of the University* (1832) by Sedgwick, Buckland's counterpart at Cambridge.

Some of the biblical literalists were active and competent naturalists; in several instances they were provincial clergymen whose

---

[1] E.g. Green, *Religion at Oxford and Cambridge*, 1964, 308–9. But see Page, 'Diluvialism and its Critics', *Toward a History of Geology*, ed. Schneer, 257–71.

[2] Compare Millhauser, 'The Scriptural Geologists', *Osiris* xi (1954), 65–86.

knowledge of the geology of their particular district was second to none, but whose theory of geology was traditional and as old-fashioned as the education they had received in their university days. Competent local observations in the context of traditional theory went back as far as Catcott and his work in the Bristol area.[3] Among the several naturalists to continue in this line during the early nineteenth century was Joseph Sutcliffe, a Methodist clergyman from the Bath area. In his *Short Introduction to the Study of Geology* (1817) he intended to vindicate 'The Mosaic Account of the Creation and the Deluge', arguing, among other things, that stratified rock formations, such as those in which the coal beds occur, had been deposited by the tides of the deluge. In this system of geology the limestone formations in which the various bone caves were discovered had also been deposited by the waters of the deluge.[4]

Early disagreement with Buckland's cave theory came from two Yorkshire clergymen, each interested in local history and knowledgeable about geology. If one was to explain Kirkdale Cave and retain the notion that the limestone formation it occurs in had been formed during the deluge, there were two possibilities; one must suppose either that the teeth and bones had been washed into chalk holes already formed during the deluge, or that the caves and their fauna were postdiluvial, not antediluvial as Buckland suggested. The first of these alternatives was championed by George Young, a Presbyterian clergyman from Whitby; the second by William Eastmead, a Congregationalist minister from Kirby Moorside. Young was a respected naturalist who wrote on the environs of his town, and whose *Geological Survey of the Yorkshire Coast* became an important source for Yorkshire geology. Its first edition (1822) contained a description of Kirkdale Cave, but it was in the second edition that Young attacked Buckland's hyena den theory (1828).

Young's description, and the theory which he put up as an alternative to Buckland's, also formed the substance of two lectures read before the Wernerian Natural History Society (1822). In reference to the hyena den theory he said:

The opinion which I hold, in opposition to this theory, and which is already published in the Geological Survey of the Yorkshire Coast, is, that, as

---

[3] Neve and Porter, 'Alexander Catcott: Glory and Geology', *Brit. Journ. Hist. Sci.* x (1977), 37–60.

[4] See also Sutcliffe's *Refutation of Prominent Errors in the Wernerian System of Geology*, 1819, and his *Geology of the Avon*, 1822.

immense numbers of animals of all descriptions were drowned by the Deluge, vast masses of animal matter must have been floated or drifted about in all directions, and quantities of this matter descending to the bottom, while the diluvian waters yet covered the present strata, might be drifted into such chasms or fissures of rocks as were then open, great part of which might be subsequently covered up by the deposition of the alluvial beds at the final retiring of the waters; and that, as the bones and flesh of the animals, by being tossed about, would be broken, mangled, and mixed in wild confusion, the accumulation of such mixed relics as were found in Kirkdale Cavern may thus be accounted for.[5]

Eastmead, soon after the discovery of Kirkdale Cave, wrote an essay on it, but in keeping with his theory that the caves were post-diluvial, and thus contemporary with modern man, he included this essay in a book on the history of Kirby Moorside, *Historia Rievallensis* (1824). Eastmead did not question the hyena den theory. 'That this Cave was the habitation of Hyaenas,' he wrote, 'is, I conceive, from the facts and appearances connected with it, almost beyond a doubt.'[6] But because he believed that the limestone in which the cave existed had been deposited during the deluge, he argued that the date of habitation by hyenas had to be postdiluvial. He reasoned that after the deluge hyenas and the animal species they preyed on had come to inhabit England, just as wolves once inhabited this country. Their eventual disappearance he attributed to the fact that the hyenas had rendered their prey extinct, and in the end had eaten one another till at last the only surviving hyena had devoured itself.

These two theories, opposed to Buckland's antediluvial hyena den reconstruction, were not confined to Yorkshire authors. Two other writers, commanding a national readership, came out in their support; Granville Penn advocated the drift theory, and George Bugg the postdiluvial alternative. Each wrote a voluminous work on the connection between geology and the Bible. Penn published his influential *Comparative Estimate of the Mineral and Mosaical Geologies* in 1822, in which he argued, as had Deluc, that at the time of the deluge land and sea had changed places. No discussion of Kirkdale Cave was included, but as soon as Penn had gained knowledge of Buckland's cave work he published a lengthy *Supplement to the*

[5] 'On the Fossil Remains of Quadrupeds', read 30 Nov. 1822, *Mem. Wernerian Nat. Hist. Soc.* vi (1832), 172.
[6] 30.

*Comparative Estimate* (1823), chiefly to attack the hyena den theory. He strongly argued for the drift alternative and expressed the conviction that the cave fossils were the remnants of animals drowned in the waters of the deluge and swept from tropical regions to Yorkshire, where they became enclosed in the still soft and freshly deposited limestone formation.

Penn added a variant of his own to the theory by suggesting that the cave had originated around the enclosed carcasses as a result of their putrefaction which had produced gases and distended the space around the entombed corpses. Penn recognized the element of incongruity between the first and the second part of Buckland's *Reliquiae Diluvianae*: 'The *Reliquiae Diluvianae* has, indeed, ably and unanswerably added to the demonstrations of the truth of the *sacred history of a deluge*; not by hypotheses of *hyaenas' dens* or *bears' dens*, but, by its sagacious discrimination between *alluvial* and *diluvial* productions, duly limiting the operation of the former, and vindicating to the latter its own proper and exclusive effects . . .'[7]

Bugg wrote his *Scriptural Geology* (1826–7) expressly in refutation of the geology of both Cuvier and Buckland. His treatment of the subject, in particular the implications of modern geology for biblical exegesis, was lucid and perceptive. Like Davy (though disapprovingly) he recognized that the significance of Buckland's cave work consisted in its corroboration of Cuvier's theory of a succession of periods of geological history by giving a vivid portrayal of the last of previous worlds: 'The reader must particularly understand that if Dr. Buckland's Theory of the Caves be just, it will prove the material points of *M. Cuvier's* "Theory of the Earth" to be correct. For, if these *Caves* occupied their place in these limestone rocks before the Deluge, it is evident that there must have been one or more *revolutions before* the Flood.'[8]

Bugg further argued, as had Eastmead, that the limestone rocks had been deposited at the time of the deluge, and because the caves and their fossil content were of a later period, that they had to be postdiluvial. Bugg mistrusted the novel skill of comparative anatomy and he believed that the cave fauna had not become extinct. Instead, he argued that those features used to identify extinct species could in fact have been produced by variations of

[7] *Supplement*, 46. The putrefaction theory was anticipated by Esper, *Ausführliche Nachricht*, 1774, 104–5.
[8] ii, 187–8.

climate, food, or change of place over a period of four or five thousand years.

Several other biblical literalists continued to oppose Buckland's cave work. The most influential of these was probably George Fairholme. In two books and in contributions to such periodicals as the *Philosophical Magazine* and the *Christian Observer* he argued that the biblical deluge had been unique, the only one of its kind, and the cause of the deposition of nearly all fossiliferous rocks. Fairholme rejected Buckland's cave theory, because he believed that dry land and the sea bed had changed places during the deluge, and *eo ipso* no antediluvial caves could exist on today's continents.[9]

An early division of the rocks which compose the earth's crust recognized Primary, Transition, Secondary, and Tertiary deposits. The Transition and Secondary formations were the extensive, fossiliferous rocks between the Primary granite basement and the relatively restricted Tertiary strata of the London and Paris basins. Biblical literalists generally ascribed the Transition and Secondary rocks to the action of the deluge. A variant on this theory, slightly accommodating to modern geology, was that these rocks had formed at the floor of the antediluvial ocean during the 1,500 or 2,000 years which had elapsed since the year of creation. This theory, which derived much from Deluc's scheme, required that at the time of the deluge land and sea had been interchanged. Andrew Ure, a chemist who taught at the Andersonian University, advocated this view in his *New System of Geology* (1829). However, he did not disagree with Buckland's cave theory and, to accommodate it, preserved some of the antediluvial dry land from the otherwise general *bouleversement* of the deluge.

Similar views were expressed by Sharon Turner and George Croly. Turner commented with respect to the antediluvial period: 'This interval was at least a period of one thousand six hundred and fifty-six years; and therefore allows that space of time for all the formations between the Primordial and the Tertiary. The violent changes which occurred at the Diluvian ruin, seem to be most connected with the Tertiary Geology.'[10] Croly was of the same opinion, though he used the chronology of the Septuagint (not

---

[9] 'Some Observations of the Nature of Coal', *London and Edinburgh Phil. Mag. and Journ. Sci.* iii (1833), 245–52; 'A Layman on Scriptural Geology', *Christian Observer*, 1834, 479–92.

[10] *The Sacred History of the World*, 1832, 465.

1656, but 2256 years of antediluvial time), and he rejected Buckland's cave theory in favour of the drift alternative.[11]

Buckland was undoubtedly familiar with the criticism of his cave theory; the sale catalogue of his library lists several books by biblical literalists. However, he did not enter into polemics with his critics, except when he answered a member of the Wernerian Natural History Society who asserted that modern hyenas do not drag their prey into their dens. After a lapse of several years Buckland countered by presenting documentation about contemporary hyena dens in India which appear to be facsimiles of the antediluvial bone caves.[12] Occasionally, others came to Buckland's defence. James Smithson, of Smithsonian fame, wrote a sharp critique of Penn's *Supplement*.[13] An anonymous correspondent who signed himself 'An Admirer of Buckland' wrote disapprovingly about Bugg's *Scriptural Geology*.[14] More substantially, Samuel Wilks, editor of the *Christian Observer*, used his influence to lessen criticism of the new geology. In a number of extensive footnotes, added to a letter by Fairholme (who signed himself 'A Layman'), Wilks defended Buckland and his school using geological ammunition provided by Conybeare.[15]

However, from Edinburgh in the north to London in the south, tacit or explicit approval was given to biblical literalism by such respectable establishments as Jameson's Wernerian Natural History Society and the Royal Institution of Great Britain. The Wernerian Society provided at its meetings and in its *Memoirs* a forum for the criticism of Buckland's cave theory. William Brande, in his geology lectures at the Royal Institution, presented Penn's drift theory as a valid alternative to Buckland's hyena den theory.[16] Reviews of Penn's work in the early 1820s and Fairholme's in the middle 1830s commended their books. The *Quarterly Journal* of the Royal Institution published a lengthy and favourable review of Penn's *Comparative Estimate*.[17] Of Fairholme's *General View of the*

---

[11] *Divine Providence*, 1834, 104 ff.

[12] Knox, 'Notice Relative to the Habits of the Hyena of Southern Africa', *Mem. Wernerian Nat. Hist. Soc.* iv (1822), 383–5. 'Letter of Professor Buckland to Professor Jameson, and of Captain Sykes to Professor Buckland, on the Interior of the Dens of Living Hyaenas', *Edinburgh New Phil. Journ.* ii (1827), 377–80.

[13] *Ann. Phil.* viii (1824), 50–60.

[14] *Mag. Nat. Hist.* ii (1829), 108–9.

[15] See Conybeare to Buckland, 4 Aug. 1834, NMW, Bu P.

[16] *Outlines of Geology*, 1829, 78.

[17] Anon., 'Mineral and Mosaical Geologies', *Quart. Journ. Sci.* xv (1823), 108–27.

*Geology of Scripture* the *Athenaeum* felt 'that these pages contain a great deal that is extremely deserving the serious consideration of the geologist'.[18] The degree of clerical approval may be measured by the fact that almost half of the *c.* 200 subscribers to Bugg's *Scriptural Geology* were clergymen.

It is impossible to find among the biblical literalists a common denominator of class, region, religious denomination, or political party. What probably comes closest to a common denominator was a belief in the value of written documents as evidence in questions about the past: written documents as opposed to natural objects such as rocks and fossils. This belief stemmed from an educational tradition in which classical scholarship reigned supreme. Questions about the history of the world, its chronology, its periodization, even its major physical vicissitudes (such as a calamitous inundation) were to be answered first and foremost from a study of such written documents, the most reliable of which was believed to be the Bible. Not just the past, but also the future history of the world could be discovered by a careful exegesis of books of prophecy such as Daniel and the Revelation of St. John. This belief was in turn connected with millenarian anticipations.

Much biblical literalism was written in the context of a tradition of learning in which geological facts were secondary to written evidence. Penn asserted with Buckland in mind that 'the eminent Professor of Mineralogy concedes too much to the authority of the *phenomena*, and too little to the authority of history; too much to the *numerous revolutions* adventurously propounded by Cuvier, and too little to the *binary revolutions*, lucidly indicated and distinctly limited by Moses'.[19] Fairholme too emphasized that the early history of the world could not be reliably reconstructed from scientific evidence alone. The title of Turner's book, *The Sacred History of the World*, implied the same. Both Penn and Croly were competent classicists and had shown an interest in the exegesis of prophecy; Penn in his *Christian's Survey of all the Primary Events and Periods of the World* (1811 and several later editions) in which he dealt with questions of millenarianism, and Croly in his *The Apocalypse of St. John* (1827).

The connection with this tradition of learning put biblical literalism in close and immediate contact with Oxford and

[18] Anon., *A General View of the Geology of Scripture* by George Fairholme', *Athenaeum*, 13 Apr. 1833, 228.

[19] *Supplement*, 119–20.

Cambridge universities. Penn and Croly, for example, had received their education at Oxford. Frederick Nolan, another Oxford graduate, brought the subject of biblical literalism inside his *alma mater*. Nolan had a considerable reputation as a classicist and a theologian, and he enjoyed the honour of having been invited to give the Boyle Lecture (1814), the Bampton Lecture (1833), and the Warburtonian Lecture (1833–6). In his Bampton Lecture on the *Analogy of Revelation and Science* Nolan attacked modern geology, and, without mentioning Buckland's name, the hyena den theory. He reiterated that Transition and Secondary rocks had been deposited by the deluge, that carcasses of drowned animals had been swept from equatorial regions to the north, and that thus a mixture of carnivorous and herbivorous animal corpses had become entombed.

From this close juxtaposition of Buckland's modern geology and Nolan's literalism within the confines of Oxford, it must be clear how far removed Buckland was from representing fundamentalism. While Nolan's Bampton Lecture was still in progress, Mary Buckland wrote to Whewell with scathing sarcasm:

By way of encouragement to my husband's labours, we have had the Bampton Lecturer holding forth in St. Mary's against all modern science (of which it need scarcely be said he is profoundly ignorant) but more particularly enlarging on the heresies and infidelities of geologists, denouncing all who assert that the world was not made in six days as obstinate unbelievers etc. etc. We have had two sermons about the Flood concerning which he has a theory, but his hearers cannot justly make out what it is, and we are to have next Sunday a sermon on the universal conflagration by which the earth is to make its exit from among the planets. Alas! my poor husband—Could he be carried back a century, fire and faggot would have been his fate, and I daresay our Bampton Lecturer would have thought it his duty to assist at such an "Auto da Fe".[20]

The defence of modern geology was taken up by Charles Daubeny in an anonymous review of the Bampton Lecture in the *Literary Gazette*. A sharp exchange with Nolan followed in which he mistook the anonymous reviewer for Buckland. The tension between geological pursuits and the tradition of classical learning came to the fore in Daubeny's review when he suggested that Nolan's lectures should more appropriately have provided a corrective 'to those evils which might be imagined to be incident

[20] 12 May 1833, TC, Wh P.

upon the many substantial encouragements given in the University of Oxford to the pursuits of classical literature',[21] rather than to those of science.

This clash between modern geology and biblical literalism, the latter seen as an offshoot of the tradition of classical learning, was not restricted to Oxford. Sedgwick strongly chided the literalists in his anniversary address as president of the Geological Society (1830) for using 'the records of mankind' to arrive at 'physical truth'. With ancient texts including the book of Genesis in mind, Sedgwick asserted that 'to seek for an exposition of the phaenomena of the natural world among the records of the moral destinies of mankind, would be as unwise, as to look for rules of moral government among the laws of chemical combination.'[22]

Baden Powell, Oxford's professor of geometry and astronomy, approvingly quoted Sedgwick in his discourse on *Revelation and Science* (1833), and he joined Daubeny in a sharp attack on Nolan and on several other biblical literalists who had written their 'Mosaical' or 'Scriptural' geologies.

By around 1830 modern geology had gained enough confidence to assert itself fairly aggressively as a branch of scientific knowledge with authority, over and against classical knowledge, to speak out about the early history of the world. However, at the time when Buckland began to establish geology at Oxford, some ten to fifteen years earlier, it had yet to acquire such confidence. This leads us back to the question of Buckland's motivation for promoting geology in the form of diluvial geology, in spite of the anomaly represented by his hyena den theory.

---

[21] 'Apology for British Science', *London Literary Gazette,* 1833, 770.
[22] Anniversary address of 1830, *Proc. Geol. Soc.* i (1834), 207.

# 4
# Geology and Classical Learning

## WORLD HISTORY

Buckland's motivation in advocating his diluvial theory has invariably been made out to be merely a matter of obscurantist religious commitment.[1] However, if we are to understand Buckland's motives, we must examine his theory within the specific context of Oxford University, the institution where Buckland received his education, and of which he was a loyal member. In the 1810s and early 1820s when diluvialism was formulated, the positions within 'the university, both of the subject of geology and of Buckland himself, were still insecure. The readership of geology was established only in 1818, and until 1825, the year Buckland was made canon of Christ Church, his financial resources were entirely inadequate. Buckland's diluvialism must be examined in the light of the audience to which it was directed. His lectures, his inaugural address, even his *Reliquiae Diluvianae*, were all addressed in the first instance to members of Oxford University. The pressure of patronage, the criteria by which his *alma mater* judged the academic credibility of geology or by which he himself would be judged worthy of a secure college position–these factors are relevant to Buckland's early enthusiasm for diluvial geology. To make himself understood, believed, and respected, he had to present his subject in a form which matched the established tradition of learning, its cognitive content, its hierarchy of subjects, and its educational purpose. Religious commitment was behind Buckland's diluvialism only to the extent that religion was an integral part of Oxford's tradition of academic learning. That tradition, in its accumulated complexity, conditioned early diluvial geology in England.

Geology, the scientific study of the history of the earth, overlapped with a long tradition of scholarship represented by

[1] See ch. 1, n. 35. Also Cunningham, *Revolution in Landscape Science*, 1977, ch. 6.

universal or world history. As late as the early modern period, this tradition of world history had included, in addition to human history (political, military, etc.), the physical history of the world: earth history. It could even include the history of the world as a planet: cosmogony. The implicit assumption of this tradition of world history had been that the earth existed as an abode for man, and that the earth and man were coeval. The study of earth history was therefore no different from that of human history, and both involved an examination of written documents. Historical evidence in this tradition consisted mainly of textual, testimonial evidence. The study of history, including that of the earth, meant the examination of ancient documents of historical or mythological content and the assessment of their reliability.

In this tradition the Bible was the source of much ancient history. The age, not only of man, but also of the earth, the division of its history into periods, the nature and time of its major upheavals were derived from biblical and other documentary sources. Biblical data were used as the blueprints for a range of subjects which, because of the biblical connection, would be prefixed with the word 'sacred'. Books of 'sacred history', 'sacred geography', 'sacred chronology', and a number of other 'sacred' subjects were published throughout much of the seventeenth and eighteenth centuries. A very influential example was Stillingfleet's *Origines Sacrae* (1662).

Although the Bible was a religious document, this tradition of world history was much more than just an affirmation of religious belief. The age of the earth was put at some 6,000 years, and the major events of earth history were limited to creation and deluge. But the common practice of reducing this to a derogatory mention of Archbishop Ussher's date of 4004 BC for the origin of the world fails to do justice to the scholarly weight and authority which sacred history accumulated throughout much of the seventeenth and eighteenth centuries.

The skills on which historical chronology was based were textual and historical criticism, for which a knowledge of classical and oriental languages, of ancient history, of astronomy, etc., was required. The man who, more than anyone else, used these skills to put historical chronology on a sound footing was Joseph Justus Scaliger, the giant of early modern Protestant scholarship. Scaliger left the turmoil of the religious wars in France to take a chair at

Leiden University, where his presence contributed much to the eminence which this university attained in humanistic scholarship. His *magnum opus* was a work on historical chronology, *Opus de Emendatione Temporum* (1583), in which among other things, he interpreted and integrated a large number of ancient calendar systems. To minimize the effect of corrupt or forged textual sources, Scaliger used the then-controversial rule that the earliest source, closest to the event described, is the most reliable. He also emphasized the independent value of non-biblical sources of documentary evidence.[2]

Among Scaliger's continental followers and popularizers were eminent scholars like Sethus Calvisius and Ubbo Emmius. This type of historical scholarship was an object of pride to the Protestants and one of envy to the Catholics, until among the latter the Jesuit Dionysius Petavius wrote an *Opus de Doctrina Temporum* (1627), intended to put Scaliger in the shade. Its abridged version, the *Rationarium Temporum*, went through at least fifteen editions and was reprinted as recently as 1849. In Great Britain there was popular interest throughout the seventeenth century in the subject of historical chronology, connected in part with millenarian preoccupations. John Swan's *Speculum Mundi* (1635) is a case in point. A more scholarly treatment of historical chronology was James Ussher's *Annales Veteris et Novi Testamenti* (1650–4), written with the assistance of Oxford's Thomas Lydiat.

The existence of several different textual sources of biblical genealogies contributed to a multiplication of chronological schemes. The Remonstrant scholar Isaac Vossius, for example, son of Gerard Johannes who was professor of chronology at Leiden, suggested in his *Dissertatio de Vera Aetate Mundi* (1659) that the authentic genealogies are to be found in the Septuagint, the Greek translation of the Old Testament, rather than in the Masoretic, Hebrew version.

Throughout much of the eighteenth century, historical chronology remained a subject of interest in Great Britain. A number of new treatises appeared in which the connection between chronology and astronomy was emphasized and worked out. Apart

---

[2]See Grafton, 'Joseph Scaliger and Historical Chronology', *History and Theory* xiv (1975), 156–85; North, 'Chronology and the Age of the World', in *Cosmology, History, and Theology*, ed. Yourgrau and Breck, 1977, 307–33; Pfeiffer, *History of Classical Scholarship from 1300 to 1850*, 1976, *passim*.

from Newton's chronological work, John Kennedy's *New Method of Stating and Explaining the Scripture Chronology, upon Mosaic Astronomical Principles* (1751) is a curious example of this genre. The third edition of the *Encyclopaedia Britannica* (1788–97) discussed the subject of historical chronology entirely as a matter of textual evidence combined with astronomical principles. The most influential summary of historical chronology, through which many in the early nineteenth century were informed about the subject, was William Hales's monumental *New Analysis of Chronology* (1809; second edition 1830), in which no attention was paid to evidence for early history from geological investigations.

The reliability of biblical over profane sources of chronology and ancient history was defended by writers like Stillingfleet, whose *Origines Sacrae* became a long-standing bestseller of Oxford's Clarendon Press. Biblical history and chronology provided therefore the framework for much of ancient world history. In Great Britain several major world histories of the seventeenth and eighteenth centuries structured the early history of the world explicitly around biblical data, from Raleigh's *History of the World* (1614) to the mammoth production of the middle eighteenth century, the *Universal History from the Earliest Account of Time* (1736–68), a nineteenth-volume work written by a group of primarily Oxbridge dons such as George Sale and John Campbell. As late as the early nineteenth century, history texts such as Alexander Tytler's *Universal History* (published posthumously in 1834) adhered to traditional historical chronology.

Sacred chronology imposed its scheme not only on world history, but also on early earth histories. An influential example was John Ray's *Three Physico-Theological Discourses, Concerning* (1) *the Primitive Chaos and Creation of the World,* (2) *the General Deluge, its Causes and Effects,* (3) *the Dissolution of the World and Future Conflagration* (1693). The title speaks for itself and illustrates the preoccupation with eschatology which often went together with historical chronology. The geological speculations of Burnet, Whiston, Woodward, but also of writers such as Hutchinson and Catcott, were encompassed by the traditional historical chronology.

Sacred history had met some early criticism. In 1655 Isaac Peyrerius put forward his thesis of pre-Adamites, which said that biblical history does not represent world history, but only Jewish history; that Adam was not the first man, but the first Jew. Others too expressed doubt about the validity of biblical chronology, among them the Marquis d'Argens or Simon Tyssot de Patot in France, Charles Blount in Great Britain, and Friedrich Wilhelm Stoss in Germany. In opposition to biblical chronology they advocated the chronological schemes of various oriental civilizations. World history, in these schemes, stretched over a very long period, and by some was conceived as cyclical and eternal. Criticism was also voiced by Pierre Bayle in his influential *Dictionnaire historique et critique* (1695–7), and, more elegantly, by deist thinkers like Bolingbroke or Voltaire. Such criticism contributed to a form of historical scepticism which denigrated the validity of historical, textual evidence.[3]

This disagreement on the validity of textual evidence was part of the antagonism between the Plutonists and the Neptunists of the late eighteenth and early nineteenth centuries. The Plutonist Hutton, for example, in line with deist prejudice, depicted earth history as an essentially cyclical and indeterminate process; his investigations showed 'no vestige of a beginning,—no prospect of an end'.[4] Several Neptunists, on the other hand, still derived their scheme of earth history from the tradition of sacred history, and they regarded textual evidence as a valid supplement to the evidence from geological observations. The chemist and convert to Anglicanism Richard Kirwan, whose *Geological Essays* (1799) stated the case for Neptunism, argued specifically for the utility of testimonial evidence:

In effect, past geological facts being of an historical nature, all attempts to deduce a complete knowledge of them merely from their still subsisting consequences, to the exclusion of unexceptionable testimony, must be deemed as absurd as that of deducing the history of ancient Rome solely from the medals or other monuments of antiquity it still exhibits, or the scattered ruins of its empire, to the exclusion of a Livy, a Sallust, or a Tacitus.[5]

At Oxford, in the early nineteenth century, the slant of geological opinion, the little there existed of it, was decidedly towards Neptunism. Edward Nares befriended Jean André Deluc and criticized deist eternalism in his *View of the Evidences of Christianity* (1805).

[3] Meyer, 'The Age of the World', 1951, *passim*.
[4] 'Theory of the Earth', *Trans. Roy. Soc. Edinburgh*, i (1788), 304.
[5] 4–5.

Among the publications which kept the tradition of sacred history alive were those of Faber; educated at Oxford, he held a number of ecclesiastical positions before becoming master of Sherburn Hospital. He followed in the footsteps of men like Stillingfleet and, particularly, Jacob Bryant, whose *New System, or, an Analysis of Ancient Mythology* (1774) had added to the testimonial evidence in support of biblical history by providing a collection of deluge sagas. In his Bampton Lecture of 1801, *Horae Mosaicae*, and in his later work, Faber argued against deist historical scepticism and in support of the veracity of sacred history. He used classical and Nordic mythology and other profane sources to argue for the historicity of the biblical deluge.[6]

To those who were steeped in the classical tradition it seemed inappropriate to transfer the authority of primary sources from textual and testimonial evidence to what seemed to be broken bits of muddy rocks and fossils crowding some small museum or a don's room. To some it seemed not illogical to interpret even the fossil assemblage of Kirkdale Cave in terms of human history, the Romans supposedly having used the cave to discard the remnants of animal fights.

A bibliographical illustration of the primacy of classical and historical scholarship at Oxford in the early nineteenth century is provided by the list of books printed at the Clarendon Press for the University, published in the *University Calendar*. Most of these books were classical texts (Aristotle, Cicero, Euripides, Homer, Plutarch, Sophocles, Xenophon, and many others, including Strabo whose *Rerum Geographicorum Libri XVII* enjoyed an 1807 edition). Books on history and theology included such examples from the sacred history tradition as Raleigh's *History of the World* (new edition 1829), Stillingfleet's *Origines Sacrae* (new editions 1817 and 1837), and Samuel Shuckford's *Sacred and Profane History of the World Connected* (new edition 1810). During the entire period of Buckland's teaching career at Oxford no book on natural history was published by the Clarendon Press, except for Martin Lister's book on shells, *Historia sive Synopsis Methodica Conchyliorum* (new edition 1826).

Further bibliographical evidence comes from the private libraries of Buckland and his circle, as indicated by sales catalogues of their libraries. Buckland, and more so someone as Conybeare, owned a substantial number of these classical and historical books; Buckland

---

[6] See his *Difficulties of Infidelity*, 1824. See also Prichard, *Analysis of Egyptian Mythology*, 1819.

possessed, for example, no fewer than three editions of Stillingfleet's *Origines Sacrae*. Manuscript notes, taken around 1820, show that he was actually influenced by Stillingfleet's opinion on the dispersal of mankind before the deluge.[7] Buckland's familiarity with this tradition of classical and historical learning was acquired in his own education at Oxford, at the completion of which he became classics tutor at Corpus Christi College, long before he taught geology. His professorial counterpart at Cambridge, Sedgwick, obtained his first fellowship through an examination in classics and mathematics.[8]

## EARTH HISTORY

At Oxford the sovereignty of human history over the subject of earth history came under threat, not as a result of historical scepticism and Huttonian geology, but because of Cuvier's historical geology. The reactions in the wake of Kerr's translation of Cuvier's *Theory of the Earth* set the stage for Buckland's inaugural address on the place of geology in Oxford's system of education. On the Continent, during the second half of the eighteenth century, geology had to some extent moved away from sacred history. As a result of works on natural history such as Buffon's *Époques de la nature* and human history such as Herder's *Ideen zur Philosophie der Geschichte der Menschheit*, the notion had begun to take hold that human history and earth history are not coeval, but that the latter started long before the former.

For some time this notion was kept at bay in Oxford, but when Cuvier's theory, in which human history is only the last of a succession of periods of earth history, became accessible in the English language, a number of prominent scholars felt compelled to react. Among them was Sumner, the bishop of Chester, who added an appendix to his *Treatise on the Records of the Creation* (1816) in which he dealt with the challenge of historical geology. The book itself was much in the tradition of *origines sacrae*, but he admitted that the history of our planet might stretch back much further than the history of mankind. Sumner emphasized the historicity of the biblical deluge, and he continued: 'But we are not called upon to deny the possible existence of previous worlds, from the wreck of

---

[7] Deluge file, OUM. Bu P.
[8] Clark and Hughes, *Life and Letters of Sedgwick*, i (1890), 99–100.

which our globe was organized, and the ruins of which are now furnishing matter to our curiosity. The belief of their existence is indeed consistent with rational probability, and somewhat confirmed by the discoveries of astronomy, as to the plurality of worlds.'[9]

This view made the deluge a crucial event linking the periods of early earth history with the last period, that of human history. John Playfair, in his review of Cuvier's theory for the *Edinburgh Review* (1814), advocated the notion of a quiet deluge, an inundation which had left no appreciable geological traces. This idea implied a complete removal of geology from the dominion of classical and historical scholarship, and it met with a disdainful response from Sumner; he wrote to Buckland saying of the idea of a quiet deluge that 'the two words can scarcely be put together without absurdity, and we cannot suppose such an event to have taken place on our globe, without leaving traces behind it in the physical as well as the traditional history of every country'.[10] This letter was written after the notion of a quiet deluge had also appeared in a review of Thomas Gisborne's *Testimony of Natural Theology to Christianity* (1818) in the *Quarterly Review*, placing it right in Oxford's backyard; Linnaeus' dictum was approvingly quoted, 'Diluvii vestigia cerno nulla, aevi vetustissimi plurima'.[11]

This *mise-en-scène* provided Buckland with the opportunity to introduce Cuvier's geological theory into Oxford by presenting it in the form of a defence against the notion of a quiet deluge, i.e. a defence of Oxford's scholarly heritage against its Scottish detractors. At one stroke Buckland brought into Oxford a theory of geology which asserted its own authority over successive periods of earth history, but which, in the form of diluvial geology, acknowledged the validity of textual evidence for the period of human history where the two subjects overlapped. In this context emphasis on the geological reality of the deluge was not just an affirmation of religious belief, but a demonstration of academic credibility. The weight of authority of Oxford's scholarly tradition and heritage was so great that failure by geology to corroborate the deluge, an event about which so much textual and testimonial evidence had been

[9] 5th edn. 1833, 340–1.

[10] 25 May 1820, DRO, Bu P.

[11] 'I observe no remains of the deluge, but many indications of high antiquity'. Anon., 'Gisborne's *Natural Theology*', xxi (1819), 54.

collected, would simply have discredited the new subject. In other words diluvial geology was modern geology presenting its credentials to a university where humanistic learning reigned supreme. By emphasizing the geological traces of the deluge more strongly than Cuvier did, Buckland attempted to make sense of the diverse influences of his own college education, the values and priorities of his academic environment, and the results of modern geology.

As early as 1815 Buckland lectured on diluvial phenomena, the superficial gravels with remains of elephant, rhinoceros, hippopotamus, mastodon, etc. But in his inaugural address of 1819, *Vindiciae Geologicae; or the Connexion of Geology with Religion Explained*, he formally presented his subject to the university. Buckland's lecture demonstrated the adjustment of geology to Oxford's tradition of learning. It was a plea for geology to be accepted as a worthy academic subject. Extreme deference was shown to the system of classical education; Buckland spoke of the 'ingrafting (if I may so call it) of the study of the new and curious sciences of Geology and Mineralogy, on that ancient and venerable stock of classical literature from which the English system of education has imparted to its followers a refinement of taste peculiarly their own.'[12] On the introduction of geology and related subjects into Oxford, he commented:

For some years past, these newly created sciences have formed a leading subject of education in most Universities on the continent, and a competent knowledge of them is now possessed by the majority of intelligent persons in our own country; and though it might on no account be desirable to surrender a single particle of our own peculiar, and, as we think, better system of Classical Education, there seems to be no necessity for making that system an exclusive one; nor can any evil be anticipated from their being admitted to serve at least a subordinate ministry in the temple of our Academical Institutions.[13]

Buckland went on to illustrate the connection of geology with England's tradition of learning, showing points of contact with the more established sciences such as mathematics, physics, and chemistry; with the tradition of physico-theology and the argument from design; and, above all, with sacred history which postulated a

[12] 2–3.
[13] 3.

recent origin of man and the occurrence of a deluge. Buckland supported his case by citing men like Cuvier, Deluc, Newton, and Paley; he quoted Sumner extensively. In this way the diluvial theory became the linchpin by which modern geology attached itself to the carriage of the Anglican tradition of learning and the clerical purpose of an Oxford education.

From around 1820 onwards a number of private and public pronouncements were made which reflected tension between geology and humanistic learning about the nature of historical evidence. This increasingly found expression in demands for university reform, for an elevation of the status of scientific subjects in the hierarchy of university learning. Buckland's initial willingness to let his subject serve just 'a subordinate ministry' became impossible to sustain as geology developed during the 1820s. Since he was loath to bring conflict into the open, Buckland did not publish his thoughts on the question of geological versus textual evidence, but his bias towards the quality and reliability of the former is evident from a number of handwritten notes, probably dating from the early 1820s. These are worth quoting for the strikingly vivid way in which Buckland made his point:

In our youth we have feasted with the heroes of the Iliad and the Enead. But we have no other document beyond the records of the poet to prove the reality of this feast at which, in imagination, we have been present. The chisel of the sculpturer and the pen of the poet have recorded the adventures and the glories of the heroes of Thebe and Troy, and the antiquary still treads the ruins of the cities which their arms defended or destroyed. But the documents of geology record the warfare of ages antecedent to the creation of the human race, of which in their later days the geologist becomes the first and only historiographer. And the documents of his history are not sculptured imitations of marble, but they are the actual substance and bodies of the bones themselves, mineralized and converted to imperishable stone.[14]

In another note, the contrast between geological and traditional historical evidence was even more sharply formulated:

The documents of human history are subject to imperfections from which the evidences of the historian of nature are free; manuscripts and inscriptions may sometimes have been forged or may originally have recorded statements which are untrue. Coins and medals may have been fabricated

[14] Miscellaneous notes, OUM, Bu P.

to impose on the credulity of the antiquary, or to gratify the avarice of the impostor; from all these sources of deception and error, the medals of nature are especially exempt; no human act can imitate those characters which have been impressed by the finger of the creator upon the exquisite organization of remains of animal bodies which, having been entombed and converted to stone within the substance of the nascent strata, are thus as it were transferred from the animal into the mineral kingdom, and have filled the foundations of the earth with unerring records of imperishable evidence of the history of the world which was witnessed by no human eye and recorded by no human pen.[15]

Others expressed themselves in print on the question of geological versus textual and testimonial evidence, particularly those with a strong interest in education and its need for reform, e.g. Copleston, Daubeny, and Sedgwick. Copleston, in his review for the *Quarterly Review* (1823) of Buckland's *Reliquiae Diluvianae*, criticized the historical scepticism of the Huttonian system of geology, and emphasized the historical nature of the subject of geology and the need to take account of written or even oral evidence. Copleston, the redoubtable defender of classical education, asserted the authority of the written records of history, particularly for the deluge:

We contend therefore boldly, that in an inquiry into the history of the world, to reject the evidence of written records, as wholly irrelevant and undeserving of attention, is in itself *illogical* and *unphilosophical*. It is true, that to assume these records to be infallible, and above all criticism, is to prejudge the question, and to supersede all inquiry: but when the case is one of remote age, and full of difficulty, when we are compelled to compass sea and land for presumptive and circumstantial evidence, to turn a deaf ear to that volume which professes to give a direct and detailed narrative of the whole transaction, is a greater violation of the laws of sound reasoning, and is a symptom of stronger prejudice against religion, than all the annals of superstition and bigotry can furnish against true philosophy.[16]

The tension between geology and human history over the authority of textual evidence reached breaking-point in the early 1830s. A number of geologists such as Daubeny and Sedgwick, joined by among others Baden Powell, argued that the records of human history, specifically the Bible, should be understood as the records of the moral destiny of mankind, and that they do not offer

[15] Ibid.
[16] xxix (1823), 142–3.

authoritative evidence of physical truths. This view was clamorously opposed by the biblical literalists, who fell back on the scholarship of such figures as Scaliger and Petavius, or on Hales's more recent version of historical chronology.

However, the autonomy of geology and its implicit rise in rank as an academic subject was not only opposed by biblical literalists among the country clergy, but also by a number of Oxford's prominent office-holders. An example was Edward Nares, the regius professor of modern history, who discussed the question of historical evidence in his *Man, as Known to Us Theologically and Geologically* (1834). Even though Nares expressed his opinions more moderately than did the biblical literalists, several of whom he cited with approval, he felt that the writings of Moses as a form of historical evidence were of better quality than geological data. 'For, it remains to be seen, whether as an *historian*, he has not given proofs of a knowledge of things altogether supernatural, considering the circumstances in which he must have been placed; and if this be the case with him, as the *historian* of MAN, and of human concerns, it must lead to a strong presumption, that he cannot, in any instance, greatly have misled us in regard to the history of the earth.'[17]

These debates must be seen, first and foremost, as a chapter in the history of university reform, as the struggle of geology in its ascent in the academic hierarchy of subjects. Because the Bible is not only a source of evidence of ancient history, but also a religious document, the encounter between geological and biblical history was apt to be presented as a religious one, particularly for the consumption of a general audience or readership. Buckland himself did this, early on, in his petition to the Prince Regent for the establishment of a readership in geology, in the title of his inaugural lecture, or in his correspondence with friends such as Mary Cole. But this should not mislead us into thinking that the issue can be reduced to geology versus Genesis.

Further illustrations of the deference paid by geology to classical learning, and the virtual termination of this around 1830, are to be found in the titles and quotations on title-pages of geology books. In the 1810s and early 1820s book titles might still be in Latin: Buckland's *Vindiciae* and *Reliquiae* are prominent examples. Quotations on the title-page might still be from classical authors; Brande's *Descriptive Catalogue* (1816) quoted Horace; Sutcliffe's

[17] 33.

*Study of Geology* (1817) quoted Lucretius, and his *Geology of the Avon* (1822) Manilius; the *Outlines of Geology of England and Wales* by Conybeare and Phillips used a quotation from Cicero. In contrast, authors from outside the academic establishment, such as Robert Bakewell and Gideon Mantell, used quotations from contemporaries. This became the rule later when, for example, Lyell quoted from Playfair in his *Principles of Geology* (vol. i, 1830). Quotations from the Bible or from early modern scientists like Bacon and Newton did not go out of fashion and were used through the entire period of Buckland's career.

# 5
# The Popularity of the Oxford School

Throughout the 1820s Buckland's diluvial theory enjoyed great popularity. His adjustment of modern, Cuvierian geology to the traditional system of classical education at Oxford proved a tremendous success, in spite of objections by biblical literalists and those who felt apprehensive about the encroachment of geology on territory which for centuries had been under the aegis of classics, history, and theology. The diluvial theory replaced the Mosaical geology of biblical literalists, removed this to the fringes of academe, and took its place as the new orthodoxy of the 1820s. At the same time the subject of geology acquired a niche, however lowly at first, in Oxford's curriculum, and the patronage of a substantial number of Anglican clergymen, even of some Evangelicals. Although Thomas Gisborne, a non-resident member of the Clapham Sect, objected to Cuvier's theory of the earth, men like Sumner and, most powerfully, Barrington, bishop of Durham encouraged Buckland in the pursuit of modern geology. Appropriately, the *Reliquiae Diluvianae* was dedicated to Barrington.

Buckland's success at Oxford manifested itself in a large attendance at his geology lectures, which included some of the most senior members of the university (figure 6). Buckland wrote to Mary Cole on 15 February 1823: 'I have this day been occupied in lecturing to an overflowing class, amongst whom I reckon the Bishop of Oxford and four other Heads of Colleges and three Canons of Christ Church on the newly discovered caves and have puzzled them all as well as myself to account for the phenomena of the cave at Paviland.'[1] To this record the names of no fewer than six university professors could have been added.[2]

Buckland's effectiveness in adjusting modern geology to Oxford's classical tradition was recognized by his colleagues.

[1] NMW, Bu P.
[2] Edmonds and Douglas, 'William Buckland, F.R.S. (1784–1856) and an Oxford Geological Lecture, 1823', *Notes and Records Roy. Soc.* xxx (1976), 141–67.

6. Buckland lecturing in the Ashmolean Museum on 15 February 1823 to an audience of senior members of the university. (This is the mirror image of the original print restoring the left–right symmetry of the map of England and Wales.)

Conybeare maintained in the *Outlines of the Geology of England and Wales* that Oxford geology provided a convincing proof 'that the institutions of academical education are far from unfavourable to the cultivation of the physical sciences'.[3] Charles Daubeny agreed, and in his *Inaugural Lecture on the Study of Chemistry* (1823) he quoted Conybeare in part:

I fully agree with the author of the excellent work on the *Geology of England*, himself a bright example of the union of scientific with classical attainments, who remarks, that in the Oxford school of Geology, a satisfactory proof has been afforded, in opposition to the misrepresentations of shallow sciolists, that the institutions of academical education are far from unfavourable to the cultivation of the physical sciences, but on the contrary might become, under proper management, the most suitable preparation for them.[4]

John Kidd, who had preceded both Buckland and Daubeny in respectively mineralogy and chemistry, believed, in his *Introductory Lecture to a Course in Comparative Anatomy* (1824), that a new dawn of Oxford science had arrived, comparable to its seventeenth-century heyday.[5]

This dawn of science at Oxford during the early 1820s was symptomatic of a modest but continued improvement of university standards, to which the change of the examination statutes in 1801 and 1807 had given the initial impetus. Thus the classical system appeared to be no impediment to the advancement of learning, not even to that of modern science. This conclusion was drawn by one of Buckland's students, Charles Lyell, in a major survey of the state of the universities for the *Quarterly Review* (1827); Lyell praised both Buckland and Sedgwick for the excellence of their geology lectures and stated his belief that 'Our universities are called upon to make no daring inroads upon their ancient constitution—to submit to no sacrifice of existing interests.'[6] When Whewell reviewed the 'Science of the English Universities' for the *British Critic* in 1831 he concluded that only at Oxford or Cambridge could a student hope to receive a good education in modern geology.[7]

The diluvial theory was a distinctive feature of the Oxford school

[3] xlviii.
[4] 40.
[5] vii–viii.
[6] xxxvi (1827), 264.
[7] ix (1831), 74.

of geology from the late 1810s and early 1820s till the end of the decade, when Oxford geology was subsumed under the English school. Buckland's diluvial work acted as a major stimulus to English geology. Even those outside the Oxford school recognized this. Bakewell's *Introduction to Geology* (1833 edition) commented with regard to English bone caves that 'their discovery may be said to have given a new impulse to geology, both in this country and on the Continent, for which we are chiefly indebted to the enlightened and indefatigable exertions of Professor Buckland, of Oxford'.[8] The *Revue Britannique*, with its continental perspective, attributed the emergence of modern geology in England to Buckland's work: 'En Angleterre, ce n'est guère que depuis 1823, époque de la publication des grands travaux du docteur Buckland, que la science géologique a pu secouer l'espèce de démon incube qui l'écrasait.'[9]

During the 1820s further diluvial work was carried out, both on bone caves and on gravel deposits. Buckland continued to pursue his interest in caves and planned a second volume of *Reliquiae Diluvianae* in order to describe the abundance of additional discoveries, but the volume was never published. New bone caves were discovered in, for example, Devon, Kent, and North Wales. Their occurrence was a source of local pride, especially if a cave had been a hyena den similar to the Kirkdale archetype. The first geological inspection of a number of these caves was made into a ceremonial event, to which senior scientists were invited.

A cave which attracted a great deal of attention was Kent's Hole, near Torquay in Devon, the geological study of which began in 1824. It yielded a very large and rich collection of fossil teeth and bones, and Buckland believed that it had been a hyena den. Among those who contributed to the examination of this particular example were, in addition to Buckland, Henry Beeke, dean of Bristol, Featherstonhaugh, Trevelyan, and a number of country clergymen of whom John MacEnery became the best known. Kent's Hole contained human artefacts in the form of flint implements, and because of these the cave gave rise to a protracted controversy.[10]

---

[8] 521.

[9] 'In England the science of geology has been able to liberate itself from the nightmarish demon which oppressed it only since 1823, when the great works of doctor Buckland were published.' Anon., 'Des travaux et des résultats de la géologie moderne', xi (1837), 15.

[10] Kennard, 'The Early Digs in Kent's Hole', *Proc. Geol. Assoc.* lvi (1945), 156–213.

During his honeymoon in 1826 Buckland visited a number of continental caves; in France he interpreted the Cave of Lunel near Montpellier as a hyena den and thought another example, the Grotto of Osselles near Besançon, similar to the German bear caves.[11] An amusing incident took place during this journey; in Palermo Buckland visited a grotto supposed to contain the bones of Rosalia, the patron saint of the city. He immediately recognized that the bones were those of a goat rather than a woman, whereupon a scandalized clergy had the relics enclosed in a casket.[12]

Across the Atlantic, Silliman used the *American Journal of Science and Arts* to encourage its readers to explore North American limestone caves for fossil remnants, giving Buckland's rules of procedure.[13] When Featherstonhaugh returned from his European journey Silliman reported that he had brought back a valuable collection of fossils and minerals, illustrative of the geology of England. He added:

But what will be extremely interesting here, is the capital series of osseous remains of the varieties of animals found in diluvial deposits in the various caves; a branch of geology illustrated and brought to light by the genius and eloquence of that extraordinary person, Dr. Buckland; they visited in company the celebrated cave at Torquay, from whence Mr. F. brought the bones of eleven different animals: all the circumstances of this cave confirm Professor Buckland's opinions, as expressed in the Reliquiae Diluvianae, of which we gave an analysis and review in volume eight, of this Journal. On various occasions we have urged the prosecution of similar inquiries here, and we repeat that we should be glad to see the attention of our geologists roused to the importance of this subject, as we have numberless caves to explore—and bones we must find, or draw some curious conclusions from their absence.[14]

Soon, however, a sensational new discovery of bone caves was reported, not from North America but from Australia. In 1830

---

[11] 'Observations on the Bones of Hyaenas and Other Animals in the Cavern of Lunel near Montpelier', read 17 Nov. 1826, *Proc. Geol. Soc.* i (1834), 3–6; 'Account of the Discovery of a Number of Fossil Bones of Bears, in the Grotto of Osselles, or Quingey, near Besançon in France', read 20 Apr. 1827, ibid., 21–2.

[12] Gordon, *Life and Correspondence of Buckland*, 1894, 95–6.

[13] The rules occur in a letter from Buckland to Miss Talbot, 31 Dec. 1822, NMW, Bu P. In abbreviated form Silliman published them in his *Am. Journ. Sci.* xi (1826), 190–1; Jameson did the same in his *Edinburgh New Phil. Journ.* Apr.–Oct. 1827, 382–3.

[14] *Am. Journ. Sci.* xiv (1828), 197–8.

osseous caves and fissures were discovered in Wellington Valley, New South Wales. Various reports and boxes of specimens were sent to such centres of geological research as Edinburgh, London, and Oxford. The fossil remnants were almost entirely those of indigenous genera of marsupials. A local naturalist hastily concluded that the occurrences were similar to the hyena dens of England and Wales and that the discovery 'supplies us, therefore, with another convincing proof of the reality and the universality of the deluge'.[15] Featherstonhaugh enthusiastically endorsed this conclusion,[16] but Buckland poured cold water on it; an examination of the fossil remnants showed no evidence for a den theory, such as gnawed bones. He was inclined to believe instead that the Wellington Valley occurrences were similar to the bone breccias of fissures and pitfalls,[17] i.e. accidental animal graves.

Bone caves continued to be a subject of geological interest long after the popularity of the diluvial theory had declined. The hyena den theory was described in all major geology textbooks. Cave paleontology reached maturity with Owen's report on 'British Fossil Mammalia' to the British Association in 1842. A booklet on the *Caves of the Earth* (1847), published by the Religious Tract Society, described such osseous caves as Kirkdale and Torquay. In 1849, the last year of Buckland's teaching career, Sedgwick's lectures were still at their most inspired when he dealt with the story of Kirkdale Cave. One of Sedgwick's students, Richard Wilton, wrote to a friend:

I wish I could preserve for you a lecture in its integrity, but even then the vehemence of his voice and the energy of his manner would be wanting. His appearance is captivating. While gazing on this time-worn, weather-beaten face, you cannot help remembering that it is no idle speculator you are listening to, but a philosopher indeed, the friend of Cuvier and Humboldt and Buckland and other veterans of modern science: one who recounts facts from his own observations, who has himself groped in dark caves in search of wild beasts' bones, and dredged whole days, through shine and shade, in river-beds . . .[18]

[15] *Edinburgh New Phil. Journ.* Oct. 1830 to Apr. 1831, 368.

[16] *Monthly Am. Journ. Geol. Nat. Sci.* i (1831), 47.

[17] Buckland to Murchison, 4 Feb. 1832, NMW, Bu P; Buckland to Featherstonhaugh, 23 Aug. 1831, *Monthly Am. Journ. Geol. Nat. Sci.* i (1831), 278–80.

[18] Wilton to Morine, 6 Nov. 1849, SM, Se P. See also Wilton's 'Notes from Professor Sedgwick's Lectures on Geology, October Term 1848, and October Term 1849', ibid. A partially different text of the letter to Morine occurs in Young's *Richard Wilton*, 1967, 69.

Sedgwick began his scientific career as a convinced diluvialist. Under Conybeare's tutelage he was introduced to bone caves, and one of his early geological contributions consisted of a defence of Buckland's definition of diluvial deposits in two letters to the *Annals of Philosophy* (1825).[19] This defence was directed against the Scottish school of geology. The Edinburgh journals which had criticized Oxford's system of classical education were critical of the diluvial theory as well. The *Annals of Philosophy* rather than the *Edinburgh Philosophical Journal* brought diluvialism to a broad readership. Sedgwick wrote to Buckland: 'The Edinburgh *savans* are sadly in the dark on these matters.'[20] And Buckland confided to de la Beche: 'I see there are lots of small hyena-squibs in the Edinburgh journals, but Sedgwick's long eighteen pounder is more than a match for them.'[21]

New contributions to the study of diluvial gravels were made by John Phillips, amongst others. He mapped their direction of dispersal across Yorkshire and concluded that a current had swept the county in south-easterly direction. The first chapter to Phillips's *Illustrations of the Geology of Yorkshire* (1829) lucidly stated the case for diluvialism: 'Of many important facts which come under the consideration of geologists, the "Deluge" is, perhaps, the most remarkable; and it is established by such clear and positive arguments, that if any one point of natural history may be considered as proved, the deluge must be admitted to have happened, because it has left full evidence in plain characteristic effects *upon the surface of the earth*.'[22] Phillips's uncle, William Smith, agreed; in a *Synopsis of Geological Phenomena*, prepared for the meeting of the British Association of 1832 in Oxford, he lent his support to the diluvial theory.[23] Diluvialism provided many provincial naturalists with a conceptual framework for their observations. One of them, Joshua Trimmer, reported the occurrence of marine shells in supposedly diluvial deposits in North Wales more than 1,000 feet above sea level.[24]

[19] 'On the Origin of Alluvial and Diluvial Formations', *Ann. Phil.* ix (1825), 241–57; 'On Diluvial Formations', ibid. x (1825), 18–37.
[20] 12 Feb. 1825, NMW, Bu P.
[21] 7 July 1825, ibid.
[22] 16.
[23] Reproduced in Sheppard, 'William Smith: his Maps and Memoirs', *Proc. Yorkshire Geol. Soc.* xix (1917), 175.
[24] 'On the Diluvial Deposits of Caernarvonshire', read 8 June 1831, *Proc. Geol. Soc.* i (1834), 331–2.

It was assumed that the current which had produced the diluvial phenomena of Great Britain had not been a local inundation, but had affected a very much larger part of the earth's surface. This assumption added interest to studies of diluvial occurrences in other countries. Buckland's students Strangways and Trevelyan extended their observations to parts of northern Europe.[25] De la Beche described diluvial deposits from various places, including Jamaica.[26] In North America Amos Eaton enthusiastically reported diluvial evidence encountered during his geological survey of the Erie Canal.[27] Probably the most novel foreign additions to the diluvial theory were made by Buckland, based on two rock and fossil collections. One was sent from Burma where a civil servant, John Crawfurd, collected specimens during a voyage up the Irrawaddy river; the other came from the Arctic where Captain Beechey sampled the shores of Eschscholtz Bay during an expedition in search of a northern passage.[28] In the latter collection, Buckland identified the fossil remains of elephant, rhinoceros, and other quadrupeds, found in frozen mud and sand. He used this material to argue that the diluvial event had been accompanied by a sudden drop in temperature and by a climatic change, the continuation of a general trend of climatic cooling throughout geological history.

Buckland's accomplishments led to something of a personality cult. His geological knowledge became proverbial, and his personal habits as well as his discoveries and theories were made the subject of jocular verse and cartoons. More than anything else the hyena den theory inspired humour and verse. The idiosyncratic manner in which Buckland delivered his lectures, mingling geological fact with jest, invited caricature. After Buckland had presented his hyena den theory to the Geological Society at its annual dinner in 1822, Lyell wrote to Mantell: 'Buckland in his usual style, enlarged on the marvel with such a strange mixture of the humorous and the

[25] See Buckland to Cuvier, 18 July 1827, BL, Bu P.

[26] 'Notice on the Diluvium of Jamaica', *Ann. Phil.* x (1825), 54–8.

[27] 'Notices respecting Diluvial Deposits in the State of New-York and Elsewhere', *Am. Journ. Sci.* xii (1827), 17–20.

[28] 'Geological Account of a Series of Animal and Vegetable Remains and of Rocks, collected by J. Crawfurd, Esq.', *Trans. Geol. Soc.* ii (1829), 377–92; see also Buckland's contribution to Crawfurd's *Journal of an Embassy*, ii (2nd edn. 1834), 143–62. 'On the Occurrence of the Remains of Elephants, and other Quadrupeds, in the Cliffs of Frozen Mud, in Eschscholtz Bay, within Beering's Strait', in Beechey, *Narrative of a Voyage to the Pacific*, 1831, 593–612.

serious, that we could none of us discern how far he believed himself what he said.'[29] Among those who wrote verse on the subject of Buckland's cave theory were Conybeare and P. B. Duncan.[30] Buckland himself encouraged this, and his wife Mary occasionally took to composing verse. Conybeare wrote a lengthy poem on 'The Hyaena's Den at Kirkdale', accompanied by a drawing which showed Buckland crawling on all fours into Kirkdale Cave surprising some spotted hyenas, one in the act of gnawing a bone.

The most prolific versifier was Duncan; one of his poems, 'The Last British Hyaena',[31] composed in the apocalyptic style of the Romantics, contains all the main elements of the den theory:

High on a rock which o'er the rising flood
Rear'd its bleak crag, the last hyaena stood.
Last of his race, for, victims of his maw,
With fratricidal, parricidal jaw,
His rage had each contemporary slain;
Crush'd every bone, sucked marrow, spine, and brain.
Potent his jaw to crack his bony ration,
Potent his stomach as the Pot of Pappion.
Full oft, like Captain Franklin, did he prey
On bones rejected in a former day.
Ere the great flood had poured its fatal wave,
Thro' the deep windings of his Kirkdale Cave.
But now, full fast he saw th' o'erwhelming surge,
Up Cleveland's steeps the roaring billows urge.
Yet ere it rose to mix him with the rest,
Thus did he growl aloud his last bequest—
"My skull to William Buckland I bequeath."
He moaned, and ocean's wave he sank beneath.
Southward a flood from Yorkshire chanced to travel,
And rolled the monster deep in Rugby gravel.
Strange things occur, but stranger few than this—
Doubt you the facts?—His head—lies where it is.

Probably the finest piece of verse dealing with Buckland and the diluvial theory was entitled 'Facetiae Diluvianae'.[32] It related a

---

[29] 8 Feb. 1822, in Mrs Lyell, *Life, Letters, and Journals of Lyell*, i (1881), 115.

[30] A number of the poems on Buckland occur in Daubeny's collection of humorous scientific verse, MC, Da P. Part of this collection was published posthumously, *Fugitive Poems*, 1869.

[31] Ibid., 119–20.

[32] Printed pamphlet, DRO, Bu P.

fictional encounter between Buckland and Noah, in the form of a short play in two acts. The encounter provided an opportunity to juxtapose, in a humorous manner, modern geological theory in the person of Buckland and traditional, testimonial evidence in the person of Noah, who had actually been witness to the diluvial event. The dialogue was an unmistakable slight upon the utility of historical, testimonial evidence. The first act is set in the High Street in Oxford where Buckland (B.) meets a friend (F.):

B. [*in a sprightly tone*] Well met, my friend, well met; the rarest thing
   Hath happen'd to me! raptures for a King!
F. Ha! I've not seen you in such gamesome plight
   Since those Hyena's bones first saw the light;
   When, pond'ring o'er the remnants of the Flood,
   Deep, deep you plung'd in stalagmite and mud.
   Another cavern? B. Trifles light as air!
   That may not with my present joys compare.
   Conviction, strong conviction brings applause
   Already, in idea, to my cause
   [*Murmuring to himself*] The happy tidings I shall soon bestow
   On wond'ring nations!—Fortune wills it so.
F. But what hath chanc'ed? B. Why thus it is my friend:
   Saunt'ring erewhile where Cherwell's willows bend,
   And kiss, in gratitude, the streams that feed
   Their silv'ry forms, to deck the verdant mead,
   I met old Noah:—nay, nay, never wave
   Your locks in doubt; 'tis true, as Kirkdale Cave.
   I met the Patriarch:—think not I mistake,
   Or that I dream;—I was, and am awake.

Buckland assures his friend that he did meet Noah, whom he recognized by 'his muddy hue' and, 'though sore of teeth bereft', 'by the grinders time had left'. He then hurries on, keen to keep an appointment with Noah, in the hope to hear his diluvial theory confirmed by the highest historical authority. The second act is set in Christ Church Middle Walk where Buckland and Noah meet and engage in a conversation in which they are completely at cross purposes:

B. [*eagerly*] Your servant, worthy Patriarch;—just in time:
   I hope you suffer not,—our northern clime
   Is rather sharp. [*with a low bow*] Permit me;—I have quoted—

N. [*bowing with solemnity*] Mr. Professor, I'm your most devoted.
B. The deluge as the cause—N. A tranquil season
  The present, Sir. B. The cause, I say,—the reason—
N. And do you keep your health? B. Of all those banks—
N. What, always busy?—well; the world gives thanks
  To those who labour thus. B. Those banks, I say,—
N. Yes, 'tis a pleasant, a delightful day!
B. Those gravel banks, most honour'd Patriarch, which—
N. [*looking round him*] Aye, you've done right I see, to dig that ditch;
  Nothing so bad as swamps! B. Those beds of gravel;—
  [*with a low bow*] You can, kind Sir, this knotty point unravel.
  These gravel beds, I take it,—N. Have you been
  At sea, Professor? 'Tis a weary scene.
B. Are certainly diluvian, Patriarch? N. How
  I e'er got through it, is a wonder now.
  A frightful whirl of waters! B. Which have born
  These gravel beds before them; doubtless torn
  From the successive strata. Their condition
  Bespeaks them rounded by severe attrition,
  As the impetuous current rushing on,
  Took from the north it's impulse. N. [*musing*] All alone
  We rode triumphant on the watery waste;
  No living forms it's heaving surface grac'd,
  Save those preserv'd by mercy, to renew
  The things of earth, which heavenly justice slew;
  All, all engulph'd within that dread profound.
  The eye but travell'd one unvaried round
  From sea to sky. B. [*with strong expression*] These animal remains
N. [*musing*] To bounteous Heaven we rais'd the choral strains,
  By gratitude inspir'd: Heaven deign'd to hear—
B. That in the limestone cavities appear,—
N. [*musing*] And ever hears where mild devotion springs.

And so the conversation continues till Noah says farewell and
departs to visit Sedgwick in Cambridge. Buckland concludes that
Noah will not shed any light on problems of diluvial geology. Exit
Buckland, with great dignity.

# 6
## Diluvial Geology in the Fine Arts

A notable feature of the Romantic period was the proximity in which science and the fine arts stood to each other. Several of the well-known Romantics contributed to both science and art. Literature and the visual arts found much inspiration not only in nature but in contemporary scientific theory as well. Men like Wordsworth and Constable, for example, were significantly influenced by geology.[1] Even the distinctive features of diluvialism were used in the apocalyptic art of the early nineteenth century. Among the favourite apocalyptic themes were the deluge, the last man, and a variety of natural disasters. The deluge was of course a traditional biblical topic, but its early nineteenth-century portrayal was novel, influenced by contemporary diluvial geology. Diluvial phenomena were evidence of the destruction of the antediluvian world, and as such portents of the apocalyptic prophecy of the ultimate conflagration of our world.[2]

Diluvialism was grafted on to Cuvier's theory of the earth, which recognized a sequence of separate geological periods, each terminated by a cataclysmal inundation; the last of these was widely held to have been the biblical deluge. This view of earth history was used by Byron, who worked it into his *Don Juan* (1819) and his *Cain* (1821).[3] He attributed the notion expressly to Cuvier in the preface to *Cain*: 'The reader will perceive that the author has partly adopted in this poem the notion of Cuvier, that the world had been destroyed several times before the creation of man.'[4] In *Heaven and Earth* (1823) Byron dealt in some detail with the deluge and its cataclysmal

---

[1] See Cannon, *Science in Culture*, 1978, 8–9. Pointon, 'Geology and Landscape Painting in Nineteenth-Century England', *Images of the Earth*, ed. Jordanova and Porter, 1979, 84–108.

[2] See Schenk, *The Mind of the European Romantics*, 1966, 32–3.

[3] Haber, *The Age of the World*, 1959, 207–9. Steffan, *Byron's 'Don Juan'*, i (1957), 241. See also Byron to Murray, 19 Sept. 1821, *Byron's Letters and Journals*, ed. Marchand, viii (1978), 216.

[4] Steffan, *Byron's 'Cain'*, 1968, 509.

devastation of the antediluvial world. He used such specifics of diluvial geology as the occurrence of fossils in caves and fissure pitfalls. The imminent disaster of the flood is described by a chorus of spirits, issuing forth from a cave:

> We, we shall view the deep's salt sources poured
> Until one element shall do the work
>   Of all in chaos; until they
>   The creatures proud of their poor clay,
> Shall perish, and their bleached bones shall lurk
>   In caves, in dens, in clefts of mountains, where
> The deep shall follow to their latest lair;
>   Where even brutes, in their despair,
> Shall cease to prey on man and on each other . . .[5]

Shelley also used geological, diluvial imagery, for example in his *Prometheus Unbound* (1820).[6] Ruskin, even before he had come up to Oxford and fallen under the influence of Buckland, composed verse on the diluvial theme, inspired by the erratic boulders in the Chamonix region.[7] A number of minor poets too wrote on the deluge; as late as 1839 John Edmund Reade, to whom Byron served as a model, wrote a drama in twelve scenes on *The Deluge* which made use of contemporary geology.[8]

Geologists were proudly aware of the literary use of their work. In the Buckland family amusement was generated by jocular versification and imitation of, for example, Byron's *Cain*; Buckland was portrayed as engaging Lucifer in dialogue, deep inside Kirkdale Cave; to Buckland's question about what is done below, Lucifer answers that he is making:

> A drink to madden Byron's brain,
> To nonsense madder still than Cain,
> To fire mad Shelley's impious pride
> To final crisis, suicide.[9]

The cause of the deluge was believed by many to have been a comet; its near passage could have produced gigantic tidal waves,

---

[5] I. iii. 170–8.
[6] E.g. Act IV, lines 314–18.
[7] See ch. 2, n. 23.
[8] See also McHenry's *The Antediluvians,* 1839, and its lengthy review in *Blackwood's Edinburgh Mag.* xlvi (1839), 119–44.
[9] Autograph poem 'The Professor's Descent', DRO, Bu P.

engulfing the continents in powerful surges of cataclysmal punishment. This idea was particularly well suited to the apocalyptic vision of John Martin, who portrayed it in his *Deluge*. Martin's paintings and engravings impress by their sublime and apocalyptic vision. Good examples are *The Fall of Babylon* (1819), *Belshazzar's Feast* (1820), and the illustrations for Milton's *Paradise Lost* (1827).[10] In most of these, the physical sphere is given a vast and grand expanse, at the expense of the human dimension. Individuals and their agony dwindle against the background of superhuman architecture or of the near-boundless vault of nature pregnant with apocalyptic portent. None of his paintings or mezzotints excels the Romantic grandeur of *The Deluge,* with its swirling magnificence showing nature's total sovereignty over man. It was painted in 1826; in 1828 Martin produced a mezzotint version, and in 1834 another oil-on-canvas which was exhibited in Paris and awarded a gold medal by Louis-Philippe.[11]

Martin's contemporary appeal was expressed by Bulwer-Lytton in *England and the English* (1833). He called Martin 'the greatest, the most lofty, the most permanent, the most original of his country, perhaps of his age'.[12] Elaborating on this he wrote:

Look at his DELUGE—it is the most simple of his works,—it is, perhaps, also the most awful. Poussin had represented before him the dreary waste of inundation; but not the inundation of a world. With an imagination that pierces from effects to the ghastly and sublime agency, Martin gives, in the same picture, a possible solution to the phenomena he records, and in the gloomy and perturbed heaven, you see the conjunction of the sun, the moon, and a comet! I consider this the most magnificent alliance of philosophy and art of which the history of painting can boast.[13]

It was indeed true that Martin used the 'philosophy' (i.e. science) of his day. He knew such geologists as Hawkins and Mantell and he illustrated ancient landscapes for their books. Cuvier visited Martin's studio in London and apparently expressed his admiration for the conception of *The Deluge* and agreed with the cometary hypothesis of the flood.[14] In addition to the cometary mechanism, a

---

[10] See Pointon, *Milton and English Art,* 1970, *passim.* Good reproductions occur in Feaver, *The Art of John Martin,* 1975.

[11] Martin's reception in France is the subject of Seznec, *John Martin en France,* 1964.

[12] ii. 136–7.

[13] Ibid.

[14] Pendered, *John Martin, Painter,* 1923, 132–4.

number of additional features of *The Deluge* appear to have been inspired by diluvial geology (figure 7). On one side of the picture blocks the size and shape of erratic boulders are rolling off a mountain-side, while on the other side similar blocks are being transported by the tidal waves. Not only is a large cave depicted in one corner of the mezzotint, but next to it there is a den of 'ferocious animals'. Among the animals depicted are several known from cave paleontology, like elephant or 'wolve'. These and other features were described by Martin in a pamphlet which accompanied *The Deluge*.[15] Even the valley morphology seems affected by the powerful action of the tidal waves. Martin's deluge engraving could have served as an illustration of contemporary diluvial geology. In fact, his style can still be recognized in illustrations of the 'Asiatic Deluge' used in geology textbooks by Louis Figuier and by Oscar Fraas.[16]

Martin returned to the idea of the deluge, producing *The Eve of the Deluge* and *The Assuaging of the Waters* (exhibited in 1840). Another major deluge picture was painted by Francis Danby, possibly in competition with Martin; it has been suggested that Martin plagiarized the original conception from Danby.[17] The latter delayed the completion of his version, and it was first exhibited in London in 1840. Danby, like Martin, conceived the deluge as a cataclysmal event of geological proportions, and he consciously differed from Poussin's portrayal, about which he wrote: 'I have always felt it is but a poor conception for the Deluge. The destruction of the whole face of the earth does not seem about to take place; it rather seems the swelling of a little brook so much above its usual size by a heavy fall of rain that a boat and some people are carried down a waterfall.'[18] Danby's painting also received much praise; the *Literary Gazette* called it 'a picture in itself making an epoch in art'.[19]

In prose writing too, diluvial geology and in particular the hyena den theory provided a new source of imagery and metaphor. A fine example is by John Stuart Mill, who in a review of Jeremy Bentham's work for the *London and Westminster Review* (1838) compared English law to the successive strata of the earth's crust; he continued: 'every

[15] *Descriptive Catalogue of the Engraving of the Deluge*, 1828.

[16] Figuier, *La Terre avant le déluge*, 1863; Fraas, *Vor der Sündfluth!*, 1866.

[17] Greenacre, *The Bristol School of Artists*, 1973, 45–8.

[18] Quoted by Adams, *Francis Danby*, 1973, 102.

[19] 1836, 349.

7. John Martin's mezzotint *The Deluge*.

struggle which ever rent the bosom of society is apparent in the disjointed condition of the part of the field of law which covers the spot; nay, the very traps and pitfalls which one contending party set for another are still standing, and the teeth not of hyenas only but of foxes and all cunning animals are imprinted on the curious remains found in these antediluvian caves.'[20]

[20] vii (1838), 492.

# 7
## Controversy with Fluvialists

The wave of popularity which the diluvial theory enjoyed was followed, almost inevitably, by a backwash. Around 1830 dissatisfaction began to be expressed with what was believed to be an excessive interest in the subject of geology and in particular an excessive identification of geology with cave paleontology. The *Athenaeum* incongruously prefaced a review of the *Fossil Flora of Great Britain* by John Lindley and William Hutton with an apparent slight upon Buckland, his canonry at Christ Church, and his *Reliquiae Diluvianae*; the reviewer felt that for the previous ten years geology had enjoyed more than its fair share of importance and patronage, and he continued: 'A man lately got a pension or place, or some such thing, of twelve hundred a year, for collecting a cartload of dry bones, and writing a volume of conjectures about the animals they belonged to; it was really delightful to read that the creatures had teeth in their jaws, wore tails, roamed in woods or lay in dens, and then died as beasts do now.'[1]

Around this time, too, the opposition to diluvialism from within the geological community was forcefully expressed by Lyell in his *Principles of Geology* (1830–3), in which he attempted to explain geological change 'by reference to causes now in operation'. In addition, during the early 1830s, successive presidents of the Geological Society of London dissociated themselves from diluvialism. These recantations culminated in Buckland's own abandonment of the diluvial theory in his Bridgewater Treatise (1836), where he no longer attributed any of the diluvial phenomena to the biblical deluge. These retractions have been interpreted as a victory for Lyellian geology.[2] This, however, is an oversimplification based on a confusion of two different issues which were involved in the diluvial debate.

[1] 1832, 545.

[2] E.g. Haber, 'Fossils and the Idea of a Process of Time in Natural History', *Forerunners of Darwin,* ed. Glass, Temkin, and Straus, 1959, 256 ff.

One question was whether the diluvial phenomena should be explained 'by reference to causes now in operation', or by the rare event of a cataclysmal inundation. With regard to this issue, which was debated to a large extent in terms of the origin of valleys, the diluvialists stuck to their guns. The opponents of diluvialism choose their examples from central France, whereas the advocates of the diluvial theory pointed to examples in England. This geographical preference is easy to understand with the benefit of hindsight; the diluvial examples are from a region of northern Europe extensively affected by glacial activity. Diluvialists, though sobered by the apparent complexity of valley morphology, greater than previously envisaged, insisted that some examples could only be adequately explained by a débâcle of some kind. From this line of diluvialism the glacial hypothesis originated around 1840.

The other question, quite unrelated to the first, was whether this débâcle was identical with the biblical deluge or of an earlier date. The retractions related to this issue: the new theory stated that the last geological deluge had occurred long before the Mosaic flood; the latter was no longer believed to have produced observable phenomena. This redating of the last geological deluge was forced upon geologists not by any Lyellian argument for uniformity, but by the fact that in spite of a thorough search during the 1820s no human fossils had been found in diluvial deposits.

The attack by Lyell on the diluvialists was inspired by the geology of the Huttonian wing of the Scottish school. This, however, did not in the least imply that diluvialism and the Wernerian wing were at all close or that they supported each other. Much of the early criticism of the diluvial theory came from Jameson and his Wernerian Natural History Society. This accords with the thesis that the diluvial controversy must be seen against the background of the opposing Oxford and Edinburgh schools of geology, rather than that of pan-European intellectual movements of Plutonism versus Neptunism. The differences between the English and Scottish geologists overrode these wider intellectual divisions.

The use of the biblical deluge in the diluvial theory encountered opposition from two extremes. In England it came from the biblical literalists, Penn, Bugg, and others, who felt that the theory severely downgraded the significance of the deluge by restricting its geological effect to no more than superficial gravel deposits and other surface phenomena. In Scotland, however, opposition came from those who believed that the diluvial theory was yet another scheme of Mosaical geology which, like its eighteenth-century predecessors, attributed far too great a geological significance to the biblical deluge. These Scottish writers argued that geology and the Bible ought to be kept apart; that physical inquiry came under the aegis of science, and that only the moral destiny of man was the proper subject of the Bible; that therefore the biblical deluge was a subject of inquiry, not for geology, but for theology and ancient history. This argument for the separation of science and the Bible was facilitated by the Edinburgh University system in which science had, for a considerable time, enjoyed an academic status independent of the humanities. The separation was not inspired by lack of faith, but was backed by serious, exegetical arguments, namely that the biblical account of the deluge excludes a mechanism of violent tidal waves; that the story of Noah's Ark implies that all species of land animals survived; and that the deluge drowned not just animals but man as well, so that human fossils ought to occur in diluvial deposits.

These exegetical difficulties were pointed out by correspondents of London's *Philosophical Magazine* in reaction to Cuvier's *Theory of the Earth* and Buckland's inaugural lecture.[3] But more frequent and protracted were the objections to the diluvial theory published in Edinburgh. In the *Edinburgh Review* Fitton argued, as had Playfair, that the biblical deluge had been a quiet event which had left no appreciable geological traces.[4] In the middle 1820s Jameson's *Edinburgh Philosophical Journal* published a number of articles by John Fleming opposing the diluvial theory. Jameson's Wernerian Natural History Society, of which Fleming had been a member since its foundation in 1808, provided a forum for such opponents of diluvialism as Boué and Young.

Fleming, a Scottish Presbyterian minister and later professor of natural philosophy, argued in an article abridged from his *Philosophy of Zoology* (1822) that the extinction of animal species had not been caused by geological upheavals, but had happened gradually and piecemeal.[5] The following year Fleming developed this theory further in a paper which dealt specifically with the

[3] Boyd, 'On Cosmogony', *Phil. Mag. and Journ.* 1 (1817), 375–8. Anon., 'Reflections on the Noachian Deluge', ibid. lvi (1820), 10–14.

[4] 'Geology of the Deluge', xxxix (1824), 229 ff.

[5] 'On the Revolutions which have taken Place in the Animal Kingdom', *Edinburgh Phil. Journ.* viii (1823), 110–22.

diluvial cave and gravel fauna of Buckland's *Reliquiae Diluvianae*.[6] Fleming did not oppose the hyena den theory, but argued that the extinction of the cave hyena had been caused by the activity of man, the expansion of human civilization. He cited extinctions which had occurred in postdiluvial times, the most sensational of which was the disappearance of the Elk, remnants of which were known from the alluvial deposits of the Irish peat bogs. This paper was followed by 'Remarks on the Modern Strata' (1825), in which Fleming suggested that the diluvial gravel deposits themselves could also be explained in terms of gradual, piecemeal events, such as the bursting of mountain lakes.[7] The arguments from these papers were integrated and further developed in Fleming's major attack on diluvialism, 'The Geological Deluge, as interpreted by Baron Cuvier and Professor Buckland, inconsistent with the Testimony of Moses and the Phenomena of Nature' (1826).[8]

Another Scotsman who opposed diluvialism was John MacÇulloch. His *System of Geology* (1831) denigrated the subject of paleontology and in particular Buckland's cave work: 'The Caverns and their remains have recently attracted so much popular attention, under that love of the marvellous which so often loses sight of science, and of truth also, that I can refer, without difficulty, to volumes in abundance, those who consider that all Geology is comprised in such pursuit as this.'[9]

Doubts about the validity of the diluvial theory entered the Oxford school of geology during the second half of the 1820s, when Lyell, Murchison, and before them Scrope, travelled the Continent and documented examples of valleys in central France, more particularly the Auvergne, which quite decidedly were not the product of a single violent rush of water, but of slow continuous erosion by the rivers flowing through them. Scrope, though influenced by Sedgwick while a student at Cambridge, drew attention to these examples in his *Memoir on the Geology of Central France* (1827). The valleys in question had been filled in by lava flows from nearby volcanoes, but had subsequently been re-excavated, a process which had been repeated several times. This

[6]'Remarks Illustrative of the Influence of Society on the Distribution of British Animals', ibid., xi (1824), 287–305.
[7]Ibid., xii (1825), 116–27.
[8]Ibid., xiv (1825), 205–39. See also Rudwick, *Meaning of Fossils,* 1972, 171 ff.
[9]i. 448.

was undeniable proof that the formation of the valleys had not taken place in a single brief event, but intermittently over a considerable period of time. It ruled out the deluge as a cause and left as the only alternative the perennial erosion of rivers. Scrope concluded from these examples that 'it is surely incumbent on us to pause before we attribute similar excavations in other lofty tracts of country, in which, from the absence of recent volcanos, evidence of this nature is wanting, to the occurrence of unexampled and unattested catastrophes, of a purely hypothetical nature'.[10]

Lyell wrote a favourable review of Scrope's book on central France for the *Quarterly Review*,[11] and he himself, accompanied by Murchison, visited the area and supported Scrope's field observations.[12] They added a further argument; a diluvial gravel deposit was found below a volcanic deposit, and in the latter a valley had been excavated. This valley was thus of a later date than the gravel and could not be referred to the process which had emplaced this sediment. These arguments against the diluvial theory were developed by Lyell in his *Principles of Geology*. Scrope, in his turn, reviewed Lyell's work favourably for the *Quarterly Review*,[13] though neither he nor Murchison went as far as Lyell in the latter's uncompromising and extreme Huttonian stance.

Buckland generally avoided controversy in print, but he made a rare exception when he wrote a 'Reply to some Observations in Dr. Fleming's Remarks on the Distribution of British Animals' (1825).[14] He defended his position against the Scotsman's criticism by pointing out that he, just as much as Fleming, believed in the importance of present-day analogues as an aid in the interpretation of the past. This belief was, after all, the very foundation of vertebrate paleontology in general and of Buckland's hyena den theory in particular. However, Buckland rejected Fleming's a priori assumption that such analogues are sufficient; he argued that where these appeared inadequate non-actualistic forces should be inferred.

[10] *Geology and Extinct Volcanoes of Central France,* 2nd edn. 1858, 209. See also Scrope, 'On the Gradual Excavation of the Valleys in which the Meuse, the Moselle, and some other Rivers Flow', read 5 Feb. 1830, *Proc. Geol. Soc.* i (1834), 170–1.

[11] xxxvi (1827), 437–83.

[12] Lyell and Murchison, 'On the Excavation of Valleys', read 5 Dec. 1828, *Proc. Geol. Soc.* i (1834), 89–91. This paper also appeared in the *Edinburgh New Phil. Journ.* , Apr. to Oct. 1829, 15–48.

[13] xliii (1830), 411–69.

[14] *Edinburgh Phil. Journ.* xii (1825), 304–19.

The staunchest advocate of the diluvial origin of valleys was Daubeny, who had visited the volcanic district of the Auvergne before Scrope. In a letter to Jameson 'On the Diluvial Theory, and on the Origin of the Valleys of Auvergne' (1831), he defended the Oxford formula which had based the diluvial theory on combined geological and historical evidence. The evidence for a deluge from history and tradition could not, within the Oxford school, be summarily discarded. This question of human, historical evidence was, Daubeny emphasized, more than a question of religion; it was one which concerned the utility of evidence from testimonial records of history in general. The mode in which the deluge might have taken place, the causes which produced it, and similar questions might not be answerable by appeal to the Bible. 'This, however, is quite foreign from the question, whether, in balancing the rival pretensions of two scientific theories, we should be justified in throwing out of the scale the evidence derived from a fact so circumstantially related in the earliest of known records, and confirmed, in the main, by the traditions of other nations?'

To Daubeny the Bible and other historical records could be used to extend the range of individual experience and observation. Thus Lyell's 'causes now in operation' would be those observed by mankind and recorded in both ancient tradition and modern history. 'If, then, a tradition had come down to us of a comet,' Daubeny wrote, 'which, by its approach to our globe, had occasioned fearful ravages on its surface, and if observation furnished nothing to contradict, but much to confirm, the *general* truth of the report, it would imply an excess of scepticism to deny the fact, merely because centuries had passed away without any similar event having taken place.'[15] At the end of his letter Daubeny noted the irony that Scotland's Huttonian theory was advocated in England by one of Oxford's graduates, and he added that, although Lyell could hardly be counted among the alumni of the Oxford school of geology, the university at large would not disown its talented alumnus.

If Buckland was the titular head of the diluvialist party, Conybeare was the party theoretician. The latter took up the challenge of the 'fluvialists', a term coined by Conybeare to denote the opponents of the diluvial theory. With his customary confidence and, to some extent, arrogance, Conybeare dismissed Scrope as 'a goose' and denigrated some of the Scotsmen, calling them

[15] *Edinburgh New Phil. Journ.* Oct. 1830 to Apr. 1831, 205 and 217.

'donkeys';[16] he irked Lyell with the accusation that, where the *Principles of Geology* dealt with geological opinion in classical antiquity, Lyell had plagiarized Conybeare's *Outlines*.[17] More substantially, Conybeare produced a number of papers in which the changes of diluvialist opinion were charted. The first was 'On the Valley of the Thames' (1829), read to the Geological Society as part of its lively debates occasioned by the diluvial controversy.[18] The second consisted of a series of letters to the *Philosophical Magazine* (1830–1) in which Conybeare reacted to the first volume of Lyell's *Principles of Geology*.[19] A third was his major 'Report on the Progress, Actual State, and Ulterior Prospects of Geological Science' (1832), presented to the British Association at its meeting in Oxford.[20] Conybeare argued that causes now in operation are inadequate to explain the English examples of dry valleys or massive accumulations of gravel and huge erratic boulders derived from parent rock not present in the drainage basin in which they now occur. These phenomena could not be explained by fluvial activity, but indicated the reality of a past cataclysmal débâcle. Conybeare's papers marked the full subsumption of the Oxford school under the English school of geology. This involved, among other things, a de-emphasis of the importance of diluvialism as a branch of geology, and an admission that the diluvial débâcle, however real, might not have been identical to the biblical deluge. Conybeare granted: 'whether the diluvial traces we still observe geologically, be the vestiges of the Mosaic deluge, or whether that convulsion were too transient, etc., to leave such traces, is quite another question.'[21]

Similar points were made, more forcefully, by Sedgwick in his presidential addresses of 1830 and 1831. He stressed the complexity of diluvial phenomena and in addition the absence of human fossils. Sedgwick accordingly recanted his diluvial beliefs, in a characteristically forthright manner:

Having been myself a believer, and, to the best of my power, a propagator of what I now regard as a philosophical heresy, and having more than once

---

[16] Conybeare to Buckland, 29 Jan. 1830, NMW, Bu P. See also 29 Jan. 1828 and 1 Feb. 1830, ibid.

[17] Lyell, 'Reply to a Note', *Phil. Mag. Ann. Chem.* ix (1831), 1–3.

[18] *Proc. Geol. Soc.* i (1834), 145–9.

[19] viii (1830), 215–19, 359–62, 401–6; ix (1831), 19–23, 111–17, 188–97, 258–70.

[20] *Report BAAS, York 1831, Oxford 1832,* 1833, 365–414.

[21] *Phil. Mag. Ann. Chem.* ix (1831), 190.

been quoted for opinions I do not now maintain, I think it right, as one of my last acts before I quit this Chair, thus publicly to read my recantation. We ought, indeed, to have paused before we first adopted the diluvian theory, and referred all our superficial gravel to the action of the Mosaic flood. For of man, and the works of his hands, we have not yet found a single trace among the remnants of a former world entombed in these ancient deposits.[22]

Lyell's letters to Mantell on the subject of the debates at the Geological Society may mislead one into thinking that Sedgwick contributed to a victory of Lyellian geology over Conybeare's diluvialism. He wrote: 'The last discharge of Conybeare's artillery, served by the great Oxford engineer against the Fluvialists, as they are pleased to term us, drew upon them on Friday a sharp volley of musketry from all sides, and such a broadside at the finale from Sedgwick, as was enough to sink the 'Reliquiae Diluvianae' for ever, and make the second volume shy of venturing out to sea.'[23] In fact Sedgwick continued to believe in the reality of geological deluges, and in the same address which contained his diluvial recantation he strongly criticized Lyell's dogmatic reliance on 'causes now in operation'. Sedgwick charged, in reference to Lyell's training as a lawyer, that his theory was a case of special pleading.[24]

Sedgwick's retraction was followed by similar recantations from Murchison and Greenough, the next two presidents of the Geological Society.[25] However, several major studies presented to the society continued to use diluvial terminology and the notion of geological deluges, e.g. those by Buckland and de la Beche on the geology of the neighbourhood of Weymouth (read 1830),[26] by Sedgwick and Murchison on the structure of the Eastern Alps (read 1831),[27] or several years later still by such rising members of the society as Joseph Prestwich or Hugh Strickland.[28]

---

[22] Anniversary address, 18 Feb. 1831, *Proc. Geol. Soc.* i (1834), 313.

[23] 7 June 1829, Mrs Lyell, *Life, Letters, and Journals of Lyell,* i. 253. See also Wilson, *Charles Lyell,* 1972, ch. 9.

[24] *Proc. Geol. Soc.* i (1834), 303. See ch. 15, n. 24.

[25] Murchison, Anniversary address, 15 Feb. 1833, ibid., i (1834), 443. Greenough, Anniversary address, 21 Feb. 1834, ibid., ii (1938), 69–70.

[26] *Trans. Geol. Soc.* iv (1836), 44 ff.

[27] Ibid., iii (1835), 416.

[28] Prestwich, 'On the Geology of Coalbrook Dale', ibid., v (1840), 460 ff.; Strickland, 'On the Deposits of Transported Materials usually termed *Drift*', Jardine, *Memoirs of Strickland,* 1858, 90 ff.

# 8
# The Problem of Human Fossils

When Buckland publicly withdrew his support for the diluvial theory in 1836 his retraction was no more caused by Lyell's attack on diluvialism than Sedgwick's had been. To Buckland and Sedgwick alike the abandonment of diluvial geology was not a rejection of cataclysmal débâcles, but of the Mosaic deluge as an example of these. The date of the last geological deluge was put at shortly before the creation of man, and the biblical flood was reinterpreted as a quiet event. This change in the date of the last débâcle was forced on Buckland as an indirect consequence of his own hyena den theory. This proved that present-day dry land had also been antediluvial dry land; thus the remnants of antediluvial man, of his art and industry, ought to occur in diluvial deposits, mixed with the fossils of such extinct species as the cave hyena.

There were, however, no known instances of human fossils. Cuvierian theory coped with this by assuming that during the biblical deluge land and sea had changed places; any remains of antediluvial man would lie on the floor of the present oceans, out of observational reach. In Cuvier's own words, 'the bones of that destroyed human race may yet remain buried under the bottom of some actual seas; all except a small number of individuals who were destined to continue the species'.[1] Thus the most recent fossiliferous deposits on present-day dry land had not been laid down by the Mosaic deluge, but by the penultimate flood which had occurred just before the appearance of man on earth. The fact that these deposits contained fossils of extinct species seemed to confirm that they belonged to a period antecedent to the last creation, described in the book of Genesis. The story of Noah's Ark implied that all species of antediluvial land animals had been saved from the

---

[1] *Essay on the Theory of the Earth*, 1813, 131. Cuvier never rejected Deluc's inversion theory of geological deluges and incongruously juxtaposed it to Buckland's hyena den theory; see for example the sixth edition of his *Discours sur les révolutions de la surface du globe*, 1830, 290 and 358–9.

destruction of the Mosaic deluge. This Cuvierian view of recent earth history was unambiguously spelt out by Kidd in his *Geological Essay*.

The absence of human fossils was highlighted by a number of infamous misidentifications of certain specimens as fossil men. One example was a vertebrate fossil described by Johann Jakob Scheuchzer as the remnants of a human being who had witnessed the deluge, *Homo diluvii testis* (1726). This famous fossil was bought by Martinus van Marum for Teyler's Museum in Haarlem. Cuvier examined the specimen in 1811 while on a visit to The Netherlands, and he established that it did not belong to a human being, but to a giant salamander which he called *Andrias Scheuchzeri*.[2] Another example was that of human skeletons in a calcareous matrix on Guadeloupe. One of the skeletons was shipped to England, presented to the British Museum, and described to the Royal Society (1814) by Charles König, keeper of the department of natural history.[3] It turned out, however, that the calcareous mass was of recent origin, an encrustment of coral sand, and not an indication of fossilization before the current geological epoch. A further example which failed the test of critical examination was that of human remains in diluvial deposits near Köstritz in Germany, reported by Thomas Weaver in the *Annals of Philosophy* (1823).[4] Such instances helped to make any presumed discovery of human fossils an a priori improbability and, if reported, the likely result of unskilled observations.

In Cuvier's system the absence of human fossils and the evidence of a recent *bouleversement* were the two principal points of agreement between geological and biblical history. Buckland's system, however, was very different. Cuvier's and Buckland's positions have habitually been lumped together and interpreted as backward and orthodox, an example of prejudiced refusal to accept the evidence for human fossils.[5] Nothing is further from the truth. The diluvial theory needed the occurrence of fossil man; Buckland in fact postulated it in his lectures, and only joined the Cuvierian

---

[2] Cuvier, *Theory of the Earth*, 138 ff. See also Jahn, 'Some Notes on Dr. Scheuchzer and on *Homo diluvii testis*', *Toward a History of Geology*, ed. Schneer, 1969, 193–213.

[3] 'On a Fossil Skeleton from Guadeloupe', *Abstracts of the Papers, Phil. Trans.* i (1832), 487–9.

[4] v (1823), 17–43.

[5] E.g. Lyon, 'The search for Fossil Man', *Isis,* lxi (1970), 69; Porter, 'Creation and Credence', *Natural Order*, ed. Barnes and Shapin, 1979, 116.

position in the mid-1830s when convincing examples had still failed
to come to light.

Anxiety about the absence of human fossils was expressed by
James Parkinson in 1821; he wrote to Buckland that 'the omission,
in works of such authority, as well as excellence, as Dr. Kidd's
Essay and your Lecture, of the fact of the remains of the human race
having never been found either in the deposits of the earlier ages or
in the alluvium of the deluge itself, must be injurious to the progress
of knowledge and ultimately to the sacred cause which we are all
anxious to uphold.' Parkinson feared that the absence of human
fossils would lead to the conclusion 'that *man was not created until after
the deluge.*'[6]

Initially Buckland was confident that human fossils would turn
up: a note of 1819 stated: 'The examination of diluvian gravel of
Central and Southern Asia would probably lead to the discovery of
human bones which is almost the only fact now wanting in the series
of phenomena which have resulted from the Mosaic Deluge.'[7]
Buckland cited Stillingfleet in assuming that antediluvial man had
not dispersed very extensively, and that human fossils were to be
expected only near the Asiatic cradle of mankind. However, he
believed that he had indirect evidence of antediluvial man. Fossil
species of *pecora* (cattle) were known from diluvial gravel, and on the
assumption that these animals had been created for human use,
their fossil remnants corroborated the postulate of fossil man. In
abbreviated lecture notes Buckland speculated:

*No pecora till deluge gravel.* Non-existence of fossil pecora renders the
existence of man improbable, and per contra, their presence in diluvium
indicates that man existed somewhere. Pecora essential to the simplest state
of society. Were created with and for man. Spread more rapidly in the
antediluvian world than man. Checks to rapid extension of human society
either in the savage or civilized state. Stillingfleet limits antediluvian man
to Asia. Pecora overspread the antediluvian world, and found in diluvial
gravel. Man did not spread to the North of Europe or America.[8]

Buckland did not take the story of Noah's Ark in a strictly literal
sense. He reasoned that it would have been impossible for all
present-day terrestrial species to have found a place and food in the
Ark. Not only did Buckland assume that certain species had

[6] 28 Jan, 1821, OUM, Bu P.
[7] Deluge file, ibid.
[8] Ibid.

become extinct as a result of the deluge, but he actually speculated that new species had been created after the deluge. Lecture notes on the biblical flood read: 'Probably the Ark had only those animals connected with man, or those of the antediluvian races that were intended to survive the catastrophe. Of the now existing 120,000 species of animals the greater part have been probably created since the Mosaic Deluge, and we know not whether all at once or at successive periods. The Mosaic statement of Noah laying food for all animals can scarcely be applied to 120,000 species.'[9] Buckland also expressed these and similar ideas in letters to such colleagues as Fitton and Pentland.[10] They form a remarkable record of the extent to which the young Buckland ventured into rather heterodox speculations in support of his diluvial theory.

Like Cuvier, Buckland was aware that in a variety of ways human bones could become mixed with older geological deposits, and thus not in fact be contemporaneous with the animal fossils in these. Not only could successive layers be reworked by floods, but human activity, particularly burial practices, could have placed human remnants inside deposits of an earlier geological epoch. Caves, especially, might be expected to have been used for sepulchral purposes. It was a matter of professional prudence to suspect such processes of contamination. In his *Reliquiae Diluvianae* Buckland expounded a sepulchral theory of human cave remains; he described the discovery of part of a human skeleton and of a variety of artefacts in Goat's Hole near Paviland. Because of its reddish colour the skeleton became known as 'the Red Lady of Paviland'.[11] Buckland believed that the human corpse had been buried inside Goat's Hole, possibly as recently as Roman times. He concluded that no genuine instances were yet known of antediluvian human remains in cave or gravel deposits.[12]

The absence of human fossils remained the Achilles heel of the diluvial theory throughout the 1820s, even though Fleming,

[9] Ibid.
[10] Draft of letter to Fitton, Jan. 1820, and to Pentland, June 1821, ibid.
[11] See North, 'Paviland Cave, the "Red Lady", the Deluge, and William Buckland', *Ann. Sci.* v (1942), 91–128.
[12] *Reliquiae Diluvianae*, 1823, 164–70. Buckland's role in the archaeology of Kent's Hole was discussed by Howarth, 'The Origin and Progress of the Modern Theory of the Antiquity of Man', *Geol. Mag.* ix (1902), 16–27. A rejoinder was written by Hunt, 'On Kent's Cavern with Reference to Buckland and his Detractors', ibid., 114–18. See also ch. 5, n. 10.

Scrope, and others drew attention to other controversial aspects of the theory. The diluvial controversy put Buckland in a very difficult position; he did believe that the fluvialists were, to a large extent, wrong. Ardent followers of Buckland looked to him for a forceful reply. Featherstonhaugh, for example, with whom Buckland conducted a regular and frank correspondence, expressed his anxiety in 1829: 'With Murchison denying the Deluge, and Brongniart giving it a soubriquet, I look to you as a good geologist, and a churchman and an Englishman to put some order to all this.'[13] But Buckland was not to be drawn into the fray. Instead he relied on the possibility that a discovery of fossil man would still be made to tip the scales of evidence in favour of diluvialism. Buckland answered Featherstonhaugh:

There is one discovery of infinite importance to my views of the diluvial question just published by Mr. Christol of Montpellier. He has read at the Académie des Sciences a notice on human bones in two caves near Sommières under circumstances which seem to show that man were coeval with the hyenas and elephants. The cave is described with much modesty and caution and with a full conviction of its importance. I know Christol personally and worked with him two or three days at Montpellier in the Cave of Lunel and on the surface and I know him to be an accurate observer and fully up to the nature of deposits in caverns. You will see his report and judge for youself. It seems to me a case not to be explained but by admitting what, if it can be established, is the most important geological discovery that I can ever hope to witness.[14]

From the late 1820s to the middle 1830s the question of fossil man was debated with new vigour as a result of discoveries in the south of France where Buckland had initiated modern cave research during his honeymoon in 1826. The most sensational finds were made in caves at Bize (département of Ande), described by Paul Tournal and Marcel de Serres (1828), and in caves at Sommières (département of Garde), described by Jules de Christol, later professor of geology at Dijon.[15] A number of other discoveries were made in England

[13] Featherstonhaugh to Buckland, 27 June 1829, RS, Bu P.

[14] 4 Nov. 1829, CUL, Se P. See also Buckland, 'Antediluvian Human Remains', *Am. Journ. Sci.* xviii (1830), 393–4.

[15] The new discoveries came to the British public through the *Edinburgh New Philosophical Journal*, e.g. 'Fossil Antediluvian Animals Mingled with Human Remains in the Caves of Bize', Apr. to Oct. 1829, 159–61; 'Caves Containing Human Remains', ibid., 368–9; see also 'Notice of New Bone Caves', ibid., 350–1. See n. 5, Lyon.

where MacEnery had continued to work in the area around Torquay.

The great interest of the subject was reflected in the fact that the Hollandsche Maatschappij der Wetenschappen in 1831 offered a prize for the best essay on the question of caves and their fossil content, a question drafted by Martinus van Marum. An essay by Serres received the society's prize and gold medal.[16] Serres described instances of human cave remains which he interpreted as coeval with the extinct cave fauna. The same interpretation was given by Philippe-Charles Schmerling to human remains in caves at Engis near Liège.[17]

These discoveries excited bipartisan interest. To Buckland and his school the occurrence of antediluvial human remains in caves would provide the missing link between geological and biblical history. To the opponents of the diluvial theory the early presence of man on earth would provide a mechanism for piecemeal extinction, removing the need for a cataclysmal inundation.

However, cold water was poured on the excitement by Jules Desnoyer in a paper to the Société géologique de France (1833).[18] He reconfirmed the doubts which had been expressed a decade earlier by Buckland, and argued that the discoveries in the south of France were not genuine, but the result of contamination. As causes he cited reworking of successive layers by floods and the use of caves as burial places, as places of refuge during times of persecution from the Druids to the Huguenots, and as living-quarters, particularly by the ancient Celts. Desnoyer cited the story, told by the Roman historian Florus, of inhabitants of Aquitaine having been enclosed in caves by Caesar and left to perish. He also pointed out that the ancient Celts used flint implements similar to those found in cave deposits. Desnoyer recognized three successive periods of cave inhabitation, the first when animal bones were introduced, the second when the Celts used the caves, and the third when the Romans dominated Europe.

This view, apparently cautious and professional, received general approval. In England both Buckland and Lyell agreed; each visited Schmerling in Belgium and examined his evidence, but

---

[16] 'Essai sur les cavernes à ossemens', *Natuurkundige Verhandelingen* xxii (1835), 1–222.

[17] *Recherches sur les ossemens fossiles découverts dans les cavernes de la province de Liège*, 1833–4.

[18] 'Proofs that the Human Bones and Works of Art found in Caves in the South of France, are more recent than the Antediluvian Bones in these Caves', *Edinburgh New Phil. Journ.* xvi (1834), 302–10.

concluded that it was unconvincing.[19] Buckland stuck to his sepulchral theory, in support of which he studied Celtic and Roman antiquities. His interest in such topics contributed to the establishment of the British Archaeological Association. At its first meeting in Canterbury in 1844 Buckland gave a memorable paper 'On the Contents of some Sepulchral Barrows near Canterbury'.[20]

The continued lack of convincing instances of human fossils gradually persuaded Buckland to separate the last geological deluge from the Mosaic flood. In lecture notes taken by Edward Jackson of Brasenose College in 1832 Buckland ascribed diluvial gravel to the action of a cataclysmal débâcle, but the notes continue pointedly: 'whether is Mosaic inundation or not, will not say'.[21] At this time Buckland had begun to write his Bridgewater Treatise which forced him to make up his mind about the diluvial theory and, when it was finally published in 1836, he disavowed diluvialism. He admitted that the Mosaic deluge is described in Genesis as a quiet event, and he added: 'The large preponderance of extinct species among the animals we find in caves, and in superficial deposits of diluvium, and the non-discovery of human bones along with them, afford other strong reasons for referring these species to a period anterior to the creation of man.'[22]

Buckland tried to minimize his diluvial recantation by relegating it to a footnote, but the reviewers did not let him get away with this. Scrope, who reviewed for the *Quarterly Review*, clearly took pleasure in elaborating Buckland's recantation, and Brewster in the *Edinburgh Review* emphasized that Buckland 'has abjured this doctrine as untenable'.[23]

The most strongly worded opposition to Buckland's sepulchral theory of human cave remains came from biblical literalists. Their position required that human fossils be found not just in caves and superficial gravel, but through the entire column of Secondary and Tertiary rocks. Fairholme, for example, spent much of a chapter in his *General View of the Geology of Scripture* (1833) in defence of the reality of fossil human discoveries; he even accepted the Guadeloupe examples as genuine.

---

[19] Buckland, Supplementary Notes, *Geology and Mineralogy*, i, 2nd edn 1837, 602.
[20] *Report of the Proceedings of the British Archaeological Association at Canterbury, 1844, 1845,* 106–113.
[21] 1, IGS, Bu P.
[22] *Geology and Mineralogy*, i, 95.
[23] lxv (1837), 13–14.

# 9
## Glaciation, the 'Grand Key' to Diluvial Phenomena

Around 1830 the interpretation of diluvial phenomena was a major focus of interest and the subject of an emotional debate at the Geological Society in London. During the following decade interest in Diluvium subsided and its place was taken by a preoccupation with older rocks and fossils. By around 1840 the interpretation of diluvial deposits became yet again the centre of attention and the subject of a renewed controversy. Interestingly, the line which separated the opposing views of 1840 was not the same as that of a decade earlier; now old allies turned adversaries and vice versa. Buckland was joined by Lyell (for a short while) and by such former opponents in Scotland as Fleming; but his long-term allies, even Conybeare, left him and united with Murchison, who voiced the most vociferous opposition.

This remarkable realignment, an early symptom of the disintegration of the English school in the course of the 1840s, was occasioned by the introduction of the glacial theory which interpreted the diluvial phenomena as the result of the activity of glaciers. This theory, introduced into Great Britain by Agassiz and Buckland, postulated the former existence of permanent snow, ice, and glaciers over a large part of the northern hemisphere. Ideas of land ice and of an ice age seemed too far-fetched to the majority of Buckland's colleagues, but to Buckland himself they represented the 'grand key' to various diluvial problems which had never been satisfactorily solved by the assumption of a huge tidal wave.

The most eye-catching of the diluvial phenomena were the large erratic boulders strewn across extensive regions of the north of Europe and America. During the 1830s the notion that their emplacement was the result of a cataclysmal event gave way to the iceberg theory. This attributed the dispersal of boulders to ice-floes, broken loose from the polar ice, carrying rocks, and losing these while drifting to lower latitudes propelled by winds and currents.

This theory had been suggested by J. C. W. Voigt as early as 1786 and discussed by Goethe and others.[1] The transport of boulders by icebergs was observed in both Arctic and Antarctic seas in the course of a number of polar expeditions under such captains as Parry and Ross; the drift of icebergs from Baffin Bay, down the coast of Labrador, and even into the Gulf of St. Lawrence became particularly well known and documented.[2] To account for icebergs dispersing boulders across the north of Europe and America, subsidence below sea level was assumed, followed by elevation to become dry land once again. This theory of elevation was supported by observations of the present-day rise, however slow, of the Baltic region.

Lyell promoted the iceberg theory in the third volume of his *Principles* (1833); he elaborated the theory of elevation in his Bakerian Lecture to the Royal Society (read 1834),[3] and in his anniversary address to the Geological Society (1836) Lyell put icebergs and vertical land movements together to account for erratics in the north of Europe.[4]

The iceberg theory was further developed by Murchison, in a paper on 'The gravel and alluvia of S. Wales and Siluria'[5] (1836) and in his *magnum opus* on the *Silurian System* (1839). He used the occurrence of marine fossils in the diluvial detritus to argue that during their emplacement much of northern Europe had been submerged below sea level. He visualized Great Britain as covered by water, its mountains rising up as islands, capped by snow and ice. From these elevations icebergs had originated and these, Murchison believed, had distributed boulders and other diluvial deposits across the country. In support of this theory he cited Darwin's unpublished notebooks, which contained observations on glaciers made in the course of the voyage of the *Beagle* around South America.[6]

The iceberg theory accounted for the deposition of erratic boulders by transport in ice floating on sea over submerged land;

[1] Voigt, *Drey Briefe über die Gebirgslehre*, 1786; *Goethe. Die Schriften zur Naturwissenschaft*, ed. Wolf et al., xi (1970), 226–7.

[2] The early nineteenth-century expeditions are described in Deacon, *Scientists and the Sea*, 1971, ch. 11.

[3] 'On the Proofs of a Gradual Rising of the Land in Certain Parts of Sweden', *Phil. Trans.* cxxv (1835), 1–38.

[4] Anniversary address, 19 Feb. 1836, *Proc. Geol. Soc.* ii (1838), 384. See ibid., 223.

[5] Ibid., 230–6.

[6] *Silurian System*, i (1839), 533–47.

the glacial theory, by contrast, used the movement of ice on dry land. Awareness of the transporting capacity of glaciers existed among naturalists in the eighteenth and early nineteenth centuries, especially among Alpine travellers who observed contemporary glaciers in action.[7] The idea of glacial transport of boulders was favourably received in Scotland; Playfair argued for it in his *Illustrations of the Huttonian Theory of the Earth* (1802), and later on Jameson made his *Edinburgh New Philosophical Journal* a platform for glacialism. Among its early papers on the glacial theory was the one by Esmark (1826) who observed that there were, especially in Norway, 'many proofs of the operation of immense masses of ice which have now disappeared'.[8]

However, these early anticipations of the glacial theory lacked the solid foundation of a systematic study of contemporary glaciers. Such a study was carried out by a number of Swiss naturalists, such as Jean de Charpentier, Ignace Venetz, and, in co-operation with them, Louis Agassiz.[9] These observations of active glaciers formed the basis for a step-by-step development of the glacial theory. Agassiz argued (1) that Alpine glaciers had formerly extended further down their valleys and across the Swiss plains, as indicated by moraines; (2) that the ice had also extended to the Jura Mountains where it had left erratic blocks; and (3) that the geographical distribution of erratics proved the former existence of a sheet of land ice across much of the northern hemisphere. The evidence for such an ice age was not restricted to deposits like moraines and boulders; Agassiz argued that polished or scratched surfaces and sculptured rocks which had been called *roches moutonnées* indicated the passage of massive bodies of ice.

In 1837 Agassiz outlined his glacial theory in a talk to the Helvetic Natural History Society in which he concluded that the dispersal of erratic blocks was part of 'the vast changes occasioned by the fall of the temperature of our globe previous to the commencement of our epoch'.[10] The following year Agassiz presented his theory to the Société géologique de France; he proposed 'that at a certain epoch

---

[7] E.g. G. A. Deluc in a letter to the editor, 10 Nov. 1801, *Bibliotheque Britannique* xviii (1801), 286.

[8] 'Remarks tending to Explain the Geological History of the Earth', *Edinburgh New Phil. Journ.* Oct. 1826 to Apr. 1827, 115.

[9] See Lurie, *Louis Agassiz,* 1960, *passim.*

[10] 'Upon Glaciers, Moraines, and Erratic Blocks', *Edinburgh New Phil. Journ.* xxiv (1838), 381.

the whole of Europe was covered with ice'.[11] In 1840 he published the epoch-making *Études sur les glaciers* in which he gave a detailed description of contemporary glaciers and ended with a chapter on their former existence outside the Alpine region.

The true test of the glacial theory was not, however, to be found in the Alps, in the vicinity of existing glaciers, but in northern Europe and especially in Great Britain, where Buckland and others had made a systematic and extensive study of diluvial phenomena. Moreover, the prestige of the Geological Society of London had turned it into a supreme court of geological opinion, and its verdict on the glacial theory would to a significant extent determine its viability. Agassiz's first English ally was Buckland, who in October 1838 visited Switzerland, where he examined the activity of glaciers, and, after initial resistance, accepted the glacial emplacement of erratics in the Jura Mountains.[12]

In 1840 the glacial theory was presented to the Geological Society with Buckland presiding. In June of that year a paper by Agassiz was read (*in absentia*) 'On the Polished and Striated Surfaces of the Rocks which Form the Beds of Glaciers in the Alps'. Such sculptured rock surfaces had been described long before by James Hall from the neighbourhood of Edinburgh, and attributed to abrasion exerted by large rocks propelled along by a tidal wave. Agassiz showed that such features are in fact produced by 'the action of the ice, and of the sand and fragments of stone forming the moraines which accompany it'.[13]

In September 1840 Agassiz read a paper 'On Glaciers and Boulders in Switzerland' to the British Association which met that year in Glasgow. He outlined his glacial theory and suggested 'that at a certain epoch all the north of Europe, and also the north of Asia and America, were covered with a mass of ice, in which the elephants and other mammalia found in the frozen mud and gravel of the arctic regions, were imbedded at the time of their destruction'.[14] In his conclusions Agassiz attempted to link the conventional notions of diluvial waves and of floating icebergs with his theory of continental glaciation; he commented 'that when this

[11] 'Remarks on Glaciers', ibid., xxvii (1839), 390. See also Agassiz's translation of Buckland's Bridgewater Treatise, *Geologie und Mineralogie*, i (1839), 112.

[12] Gordon, *Life and Correspondence of Buckland*, 1894, 141.

[13] *Proc. Geol. Soc.* iii (1842), 322.

[14] *Report BAAS, Glasgow 1840*, 1841, Trans. Sect., 114.

immense mass of ice began quickly to melt, the currents of water
that resulted have transported and deposited the masses of
irregularly rounded boulders and gravel that fill the bottoms of the
valleys; innumerable boulders having at the same time been transported, together with mud and gravel, upon the masses of the
glaciers then set afloat'.[15]

From the moment of Agassiz's arrival in Great Britain in 1840
events accelerated towards the climactic discussion of the glacial
theory at the Geological Society. Murchison opposed the theory; on
26 September he wrote to Sedgwick: 'Agassiz gave us a great field-
day on glaciers, and I think we shall end in having a compromise
between himself and us of the floating icebergs.'[16] On 29 October he
wrote to Egerton: 'if you have not been frost-bitten by Buckland
you have at all events had plenty of friction, scratching, and
polishing, and next year you may give us a paper on the glaciers of
Wyvis and the 'moraines' on which you sport! I intend to make
fight.'[17]

However, the glacial theory received a favourable press from the
*Athenaeum* in London and in Scotland from the *Scotsman*. After the
Glasgow meeting Agassiz and Buckland toured Scotland and the
north of England in order to find evidence of the former existence of
glaciers. This became a triumphant tour of discovery with confirmations of glaciation in many places along the itinerary,
accompanied by enthusiastic correspondence with or co-operation
from Scottish naturalists. On 4 October Buckland wrote to
Fleming: 'We have found abundant traces of glaciers around Ben
Nevis';[18] a few days later the *Scotsman* published a letter by Agassiz
on the glacial phenomena in the Ben Nevis area. Shortly afterwards, Agassiz and Buckland parted company; the latter continued
his journey past Lyell's paternal home and on 15 October he wrote
to Agassiz: 'Lyell has adopted your theory *in toto*!!! On my showing
him a beautiful cluster of moraines, within two miles of his father's
house, he instantly accepted it, as solving a host of difficulties that
have all his life embarrassed him.'[19] Agassiz went to Ireland and

[15] Ibid.
[16] Geikie, *Life of Murchison,* i (1875), 307.
[17] Ibid., 308.
[18] White, 'Announcement of Glaciation in Scotland', *Journ. Glaciology* ix (1970), 143–5.
[19] E. C. Agassiz, *Louis Agassiz,* i (1885), 309.

wrote to Buckland with equal enthusiasm on 17 October: 'Il n'y a pas de doute que les glaciers ont également couvert l'Irlande.'[20]

This field work provided the data for three papers, one by Agassiz, one by Buckland, and another by Lyell, read and discussed at the Geological Society on 4 and 18 November and 2 December. Agassiz's contribution 'On Glaciers, and the Evidence of their having once Existed in Scotland, Ireland, and England' was a rather general paper giving proofs of former glaciation; it concluded 'that great sheets of ice, resembling those now existing in Greenland, once covered all the countries in which unstratified gravel is found'.[21]

Agassiz added a novel interpretation of the so-called parallel roads of Glen Roy, three shore-line terraces, well marked in the side of the now empty glen. Darwin had just written a paper on the subject in which he suggested that the parallel terraces were formed when Scotland had been partially submerged and the sea had reached into the glen; the successive shore-lines marked stages of uplift.[22] Agassiz, however, correctly recognized that the glen had been a glacial lake, dammed up on its open side by ice; the fossil terraces in fact marked successive levels of overflow.

The papers by Buckland and Lyell were thorough and systematic descriptions of field evidence. Buckland stressed the interest of sculptured rock surfaces in the form of *roches moutonnées* which he called, most descriptively though with an un-Victorian choice of metaphor, 'mammillated rocks'.[23]

The presentation of the glacial theory was greeted with near-uniform scepticism among the leading members of the Geological Society. Greenough objected, as he objected to most generalizations. Murchison spoke with hostility and ridicule against the theory. Whewell doubted the physical possibility of the expansion of an ice sheet across regions without major mountains. The debate on the night of 18 November became heated, and even Buckland's genial manner and eloquence could not win over his opponents, although, in the words of one of those present, the gathering became

[20] 'There is no doubt that the glaciers have also covered Ireland'. OUM, Bu P.
[21] *Proc. Geol. Soc.* iii (1842), 331.
[22] 'Observations on the Parallel Roads of Glen Roy', *Phil. Trans.* cxxix (1839), 39–81.
[23] 'Memoir on the Evidences of Glaciers in Scotland and the North of England', *Proc. Geol. Soc.* iii (1842), 332–7, 345–8. Lyell, 'On the Geological Evidence of the Former Existence of Glaciers in Forfarshire', ibid., 337–45.

excited by the critical acumen and antiquarian allusions and philological lore poured forth by the learned Doctor, who, after a lengthened and fearful exposition of the doctrines and discipline of the glacial theory, concluded—not as we expected, by lowering his voice to a well-bred whisper, ''Now to,'' etc.,—but with a look and tone of triumph he pronounced upon his opponents who dared to question the orthodoxy of the scratches, and grooves, and polished surfaces of the glacial mountains (when they should come to be d----d) the pains of *eternal itch* without the privilege of scratching![24]

At the meeting of 2 December scepticism was expressed by de la Beche, Daubeny, Greenough, Murchison, Phillips, and Whewell. Even Conybeare, Buckland's forceful ally in the diluvial debate of a decade earlier, now abandoned him; he wrote: 'As to the glacial theory—you know I am a sceptic if not an infidel.'[25] In a more jocular fashion Conybeare wrote in another letter: 'Though sadly frost bitten at this moment I dont quite believe in the former geological supremacy of the Frost King—I am afraid I see reason to prove the universal prevalence of glaciers *physically impossible*'.[26] Sedgwick too refused to go along with Agassiz's theory; he wrote: 'I have read his Ice-book. It is excellent, but in the last chapter he loses his balance, and runs away with the bit in his mouth.'[27] Sedgwick's colleague, William Hopkins, who had a reputation for the use of mathematics in geology, added his voice to the criticism of glaciation.[28] Even from abroad a concerned Jean André Deluc the younger wrote to Buckland and to the Geological Society in an effort to dissuade them from following Agassiz's lead.[29] Not long afterwards Lyell too all but renounced the glacial theory, and in the

---

[24] H. B. Woodward published the notes taken by his father S. P. Woodward of the discussion of Nov. 18, 'Dr. Buckland and the Glacial Theory', *Midland Naturalist* vi (1883), 225–9; the quotation is on p. 229. The same notes, plus those of the evening of Dec. 2, were reproduced in Woodward, *History of the Geological Society*, 1908, 138–44.

[25] 3 Nov. (year not indicated), NMW, Bu P.

[26] Undated, ibid. See also North, 'Dean Conybeare, Geologist', *Report and Trans. Cardiff Naturalists' Soc.* lxvi (1933), 15–68.

[27] Sedwick to Murchison, 26 Nov. 1840, Clark and Hughes, *Life and Letters of Sedgwick*, ii (1890), 18.

[28] This reputation was in part based on his 'Researches in Physical Geology', *Trans. Cambridge Phil. Soc.* vi (1838), 1–84. His later paper 'On the Elevation and Denudation of the District of the Lakes of Cumberland and Westmoreland' advocated the theory of emplacement of erratics by icebergs and by a diluvial wave; *Proc. Geol. Soc.* iii (1842), 757–66.

[29] 28 Nov. 1839; 10 July 1842; DRO, Bu P. See also Deluc's 'On the Glaciers of the Alps', *Edinburgh New Phil. Journ.* xxviii (1840), 15–20.

second edition of his *Elements of Geology* (1841) returned to the fold of his and Murchison's iceberg theory.

For a while Buckland continued to champion glaciation. He presented a convincing account of the theory to the Ashmolean Society at Oxford in November 1840.[30] He defended glacialism in his anniversary address to the Geological Society in February 1841, and he conducted new field work in October 1841, this time in Wales where he travelled with Thomas Sopwith. He presented the results of this research in December of that year in a paper to the Geological Society in which he emphasized the evidence for glaciation formed by polished, scratched, and grooved rock surfaces (figure 8).[31]

A small following did form; correspondence expressing agreement took place with people like Maclaren and Trevelyan.[32] Some of the former adulation for Buckland's originality still showed when Sopwith drew a famous cartoon which depicted Buckland with arctic and glacial insignia, and when Duncan composed humorous verse on a conversation between Buckland and a glacial boulder.[33] However, Buckland was not cut out to be the champion of a minority cause; his emotional constitution required the approval if not the applause of his colleagues and students. In the course of 1841 he began to make room for the iceberg theory while holding on to glaciation and an ice age. In his anniversary address Buckland reconciled the two:

. . . a middle way between these two opinions will probably be found in the hypothesis, that large portions of the northern hemisphere which now enjoy a temperate climate have at no very distant time been so much colder than they are at present, that the mountains of Scotland, Cumberland, and North Wales, with great part of Scandinavia and North America, were within the limits of perpetual snow accompanied by glaciers; and that the melting of this ice and snow was accompanied by great debacles and inundations which drifted the glaciers with their load of detritus into warmer regions, where this load was deposited and re-arranged by currents at vast distances from the rocks in which it had its origin.[34]

[30] *Abstracts Proc. Ashmolean Soc.* i (1844), no. xvii, 22–4.

[31] 'On the Glacia-diluvial Phaenomena in Snowdonia', *Proc. Geol. Soc.* iii (1842), 579–84.

[32] Trevelyan to Buckland, 3 Feb. 1841, 18 Sept. 1844; see also Agassiz to Buckland, 7 Feb. 1841, 25 Apr. 1842; OUM, Bu P.

[33] Reproduced by Gordon, *Life and Correspondence of Buckland,* facing p. 145. See Sopwith to Buckland, 28 Dec. 1841, OUM, Bu P. Duncan's poem occurs in Daubeny, *Fugutive Poems,* 1869, 90–1.

[34] 19 Feb. 1841, *Proc. Geol. Soc.* iii (1842), 516.

Oct. 16. 1841.

Near the Middle of Llanberis Upper Lake.

T. Sopwith.

8. Buckland examining glacial phenomena on 16 October 1841 near Llanberis, north-east of Mt. Snowdon in Wales.

Buckland took a similar conciliatory stance in his paper on glaciation in Wales; he admitted that drifting icebergs may also be capable of producing glacial scratches by abrading the bottom of a shallow marine environment. At another meeting of the Ashmolean Society in February 1842 Buckland concluded 'that no one has yet pointed out how we are to discriminate between the grinding and polishing effects of stones in masses of ice drifted in the water, and of similar stones fixed in the ice of glaciers slowly marching upon dry land.'[35]

Buckland's conciliatory attitude did not weaken the ferocity of Murchison's opposition; he had put forward the iceberg theory in his *magnum opus* and he decided to stick to it. This was facilitated by the similarity between the iceberg mechanism and the glacial theory; both used ice masses and their transporting capacity.[36] In a report on the geology of Russia Murchison commented that ice-floes and detritus '*grating upon the bottom of a sea, may have produced the parallel striae*'.[37] In his anniversary addresses of 1842 and 1843 Murchison extensively and emphatically argued against the glacial theory.[38] For some years the iceberg theory was prominently held by English geologists; de la Beche's *Geological Observer* (1851) devoted a long section to it. Such was the doubt surrounding the glacial theory that in 1848 a rumour had to be quashed that Agassiz himself had given up his theory.[39] The *Edinburgh New Philosophical Journal* continued to carry papers on glaciation, especially from foreign authors, but indigenous contributions were largely restricted to the dynamics of ice movement, a subject studied in particular by J. D. Forbes, professor of natural philosophy at Edinburgh.[40]

Buckland's prominent advocacy of the glacial theory has puzzled historians; how could the protagonist of 'wrong' diluvialism become the English promotor of 'right' glacialism? Some sort of 'conversion', even a loss of religious faith, has been inferred to

[35]'Further Remarks on the Glacial Theory', *Abstracts Proc. Ashmolean Soc.* i (1844), no. xviii, 19.

[36]See Hansen, 'The Early History of Glacial Theory', *Journ. Glaciology* ix (1970), 135–41.

[37]Murchison and Verneuil, 'On the Northern and Central Portions of Russia in Europe', *Proc. Geol. Soc.* iii (1842), 405–8.

[38]E.g. 18 Feb. 1842, ibid., 671–87.

[39]'The Glacial Theory not Abandoned by its Author, Professor Agassiz', *Edinburgh New Phil. Journ.* xlv (1848), 366–7.

[40]See Forbes, 'An Attempt to Explain the Leading Phenomena of Glaciers', *Edinburgh New Phil. Journ.* xxxv (1843), 221–52.

account for his change of views.[41] However, Buckland's early commitment to the diluvial theory was the very cause of his later enthusiasm for glacialism. The puzzle is not Buckland's change, but the anachronistic qualifications of 'wrong' and 'right' which would prevent the one following from the other. What actually happened was that Buckland saw glaciation as the 'grand key' to the diluvial phenomena.

Intellectually the change from a diluvial to a glacial mechanism of boulder emplacement was very small indeed; in the place of Hall's tidal wave came a huge mass of frozen water. Glaciers had the advantage over a flood that they were an actualistic force which was at once better defined and had more power to produce the various 'diluvial' phenomena, especially the grooves on rock surfaces. In addition the melting of huge quantities of land ice would be a rational source of any flood that one might wish to retain as an element in the theory. But to both Agassiz and Buckland the glacial theory was more than just an alternative to the iceberg theory for the transport of erratic blocks. It led to the conclusion that there had been an ice age, a non-recurrent episode in earth history which had not only enveloped extensive regions of the northern hemisphere in land ice, but had also caused the extinction of many mammals. This aspect of the glacial theory may have been the reason why Lyell discarded it, but it increased Buckland's interest in the theory. In his anniversary address of 1841 he commented:

Thus we find, that not only the highest and northern mountain groups in the British Islands, but vast regions also of the continents of Northern Europe and of North America have been subjected to the same great physical forces, glacial and diluvial, under much colder conditions of the northern hemisphere than prevail at present; and this apparently at a time intermediate between the extinction of European and American elephants by cold, and the creation of the human race.[42]

The connection between his former diluvialism and Agassiz's glacial theory was so specific that Buckland intended to complete, at long last, the planned second volume to his *Reliquiae Diluvianae*, though under the new title of *Reliquiae Diluviales et Glaciales*. Late 1840 and early 1841 saw the beginning of this volume, of which fragments have been preserved: figures of the glacial evidences in Wales, drawn by Sopwith, were evidently intended for the new

---

[41] Davies, *The Earth in Decay*, 1969, 263. Wendt. *Before the Deluge*, 1968, 136.
[42] *Proc. Geol. Soc.* iii (1842), 514.

volume. The unpublished fragments leave no doubt about the direct and intimate connection between diluvialism and glacialism; one sheet of notes reads as follows:

Vast field of new inquiry which the introduction of the glacial period between our epoch and the Newest Tertiary open to geological inquiry. The fact of the greater part of Europe and North America having for many years been sealed up under a cover of frozen snow converted to the state of glaciers is certain. But whether this state endured 100 or 1,000 years is not yet clear. Agassiz thinks that it destroyed every living being upon the surface of the globe. Thus the flood that caused the Diluvium which in my Bridgewater Treatise I have put back to the latest of the many geological deluges, was probably due to the melting of the ice. The details of the ice flood will fill a volume and will constitute volume two of the Reliquiae Diluviales et Glaciales which for fifteen years has been retarded for lack of the grand key which Agassiz has supplied in his Études Glaciers.[43]

Buckland, however, discouraged by the very considerable opposition among his colleagues to Agassiz's theory, never completed this second volume. By the end of 1841 his enthusiasm and conviction had already withered; in response to a question about the theological implications of the glacial theory Buckland answered dispiritedly: 'The glacial theories of Agassiz are at present so much disputed and will probably require so much modification that it is at present hazardous to found any argument on his conjecture that a large portion of the Northern Hemisphere was for a very long time enveloped in a winding sheet of snow and ice.' He concluded: 'I consider the question not yet to have assumed sufficient form of consistence to warrant anything more than allusion to it as a thing quite unsettled, in any theological argument.'[44]

[43] Glacial theory file, OUM, Bu P.
[44] Buckland to Duncan, 15 Dec. 1841, ibid. For the further history of the glacial theory see Davies, *The Earth in Decay,* and Chorley *et al., History of the Study of Landforms,* 1964.

PART II

# WORLDS BEFORE MAN

## *The New Perspective of Progressive Earth History*

This world was once a fluid haze of light,
Till towards the centre set the starry tides,
And eddied into suns, that wheeling cast
The planets: then the monster, then the man;
Tattooed or woaded, winter-clad in skins,
Raw from the prime, and crushing down his mate;
As yet we find in barbarous isles, and here
Among the lowest. . . .

Tennyson, The Princess, ii. 101–8

# 10
# Strata as Pages of Earth History

## REGIONAL STRATIGRAPHY

To a casual observer it may seem that the rock masses which make up the crust of the earth are distributed in a fairly haphazard manner; no systematic relationship is immediately evident among the different rock formations seen in railway cuttings, cliff sections, or mountain exposures. The impression of irregularity is often enhanced by the dislocation, folding, or faulting of the originally horizontal strata. One of the achievements of the so-called heroic age of geology (1790–1820) was the discovery that this disorder is only apparent, and that in fact the various rock formations are systematically arranged. The position of a particular formation relative to others above and below is fixed and the same anywhere in the world.

The study of the geological occurrence of rock units is called stratigraphy; the classification table which shows the vertical succession of these units has been named the stratigraphic column. This column was, in its early version, a collation of geological sections from separate parts of Europe, correlated on the basis of superposition, composition, and, to a lesser extent, fossil content. It was to earth history what Scaliger's *De Emendatione Temporum* had been to ancient history more than two centuries earlier, namely a universal time-scale without which the separate histories of different countries and cultures could not have been interrelated to form a coherent world history. The stratigraphic column was also the *sine qua non* for an absolute chronology, the dating in numbers of years of the age of the earth and of the successive periods of its history.

Histories of geology tend to focus on the stratigraphy of the 1830s and after, on the work by Murchison and Sedgwick, or on the use of fossils to date rocks.[1] Yet the most formative phase of stratigraphy

[1] Examples range from Geikie's *Founders of Geology*, 1962, to Albritton's *Abyss of Time*, 1980.

111

was earlier; in England it occurred in the 1810s and, to a lesser extent, the 1820s, when the very possibility of a universal time-scale of earth history was still a matter of controversy.

As early as about 1820 the outline of a stratigraphic column existed, a skeleton from which the later, essentially complete version of approximately 1840 grew by gradual addition and partial correction. Buckland produced this early table of international classification and correlation. At the time several seventeenth- and eighteenth-century forerunners of stratigraphy were acknowledged. The most remarkable among them was probably Nicolaus Steno, but later names like Giovanni Arduino, Georg Christian Füchsel, Johann Gottlob Lehmann, and John Michell were also mentioned. However, the lineage of Buckland's skeleton column did not extend back much beyond the 1790s.

During this decade a school of mineralogical stratigraphy, originating with Abraham Gottlob Werner, began to spread across Europe. Werner taught at the Mining Academy of Freiberg in Saxony where he delivered an annual course of lectures from the academic year 1786–7 till the year of his death, 1817. An early version of his stratigraphy appeared under the title *Kurze Klassifikation und Beschreibung der verschiedenen Gebirgsarten* (1787). But Werner developed his ideas more fully during the 1790s in his lectures to an eager and international student audience. The list of his pupils included the names of such later luminaries of science as Leopold von Buch (1790–3), Alexander von Humboldt (1791-2), E. F. von Schlotheim (1971–3), J. F. d'Aubuisson (1797–1802), and Robert Jameson, who entered the Freiberg school in 1800. Werner's impact can be measured by the fact that at the time of his death over twenty of his pupils occupied professorial positions at institutions of higher education in Europe.[2]

The basic unit of Werner's stratigraphy was the formation, a lithological unit in the succession of rock masses. Four periods of earth history, based on such units, were recognized: (1) the Primitive (or Primary) period (crystalline rocks like granite, gneiss, and slate; no fossils); (2) the Transition period (mainly greywackes; sparse fossils); (3) the Floetz (or Secondary) period (mainly

---

[2]Gottschalk, 'Verzeichniss Derer, welche seit Eröffnung der Bergakademie auf ihr studirt haben', *Festschrift zum hundertjährigen Jubiläum der Bergakademie zu Freiberg*, 1866, 221 ff.; Wagenbreth, 'Werner-Schüler als Geologen und Bergleute', *Freiberger Forschungshefte* C 223 (1967), 163–78.

sandstones with fossils); and (4) the Newest Floetz or Alluvial period (fossiliferous gravel, sand, and clay). Volcanic rocks were classified separately. This stratigraphic system was based primarily on the rock succession in the region of Werner's residence where the older portion of non-fossiliferous crystalline rocks predominated.[3]

Werner speculated that initially, some one million years ago, the earth had been enveloped by a universal ocean. From its waters the older crystalline rocks originated by chemical precipitation. As the ocean subsided an increasing proportion of new rocks had been detrital (eroded from older deposits). The theory of a universal ocean encouraged the Wernerians to apply the fourfold classification of formations to other regions of the globe; a universal ocean was likely to have produced universal formations. 'Almost all the primitive, transition, and floetz formations, are universal deposits', Jameson declared.[4]

Among the chemical precipitates Werner also included basalt. Others believed that it was igneous in origin. This led in the course of the 1780s to a controversy between Neptunists and Vulcanists, the latter headed by Werner's own former pupil J. C. W. Voigt.[5] During the same decade James Hutton developed his Plutonist 'System of the Earth' (1785). He proposed that most rocks are not chemical but detrital in origin, eroded from the continents, deposited on the sea floor, lithified by heat from below, and subsequently uplifted to form new continents. Thus would start a new cycle of erosion, deposition, lithification, and uplift, *ad infinitum*. To Hutton not only basalt but also granite was of igneous origin. Playfair popularized (1802) Hutton's *Theory of the Earth* whereas Jameson continued to champion the Wernerian cause in his *Elements of Geognosy* (1808), the third volume of his *System of Mineralogy*. This led to an acrimonious debate between Neptunists and Plutonists and turned Edinburgh and the Scottish school during the early part of the nineteenth century into a geological battlefield.[6]

---

[3] Wagenbreth, 'Abraham Gottlob Werners System der Geologie, Petrographie und Lagerstättenlehre', ibid., 83–148. Ospovat, 'Reflections on A. G. Werner's "Kurze Klassifikation" ', *Toward a History of Geology*, ed. Schneer, 1969, 242–56.

[4] *System of Mineralogy*, iii (1808), 63.

[5] Wagenbreth, 'Abraham Gottlob Werner und der Höhepunkt des Neptunistenstreites um 1790', *Freiberger Forschungshefte* D 11 (1955), 183–241. Engelhardt, 'Neptunismus und Plutonismus', *Fortschritte der Mineralogie* lx (1982), 21–43.

[6] Ibid.

The Wernerians have been dismissed as the 'wrong' party, which lost to the Huttonians.[7] This may be true as far as the theory of the chemical origin of basalt and granite is concerned, but Werner's system was much more than a theory of chemical lithogenesis; it was a mineralogical stratigraphy which stimulated and directed historical geology, as Hutton's cyclical theory of the earth failed to do.[8] Cuvier commented that with Werner 'the most remarkable epoch of the science of the Earth commences; and we may even say, that he alone has filled that epoch'.[9]

This eulogy was to some extent Gallic hyperbole; Cuvier was all too conscious of the significance of his own contributions to the subject. Werner was first and foremost a mineralogist, and his interest was focused on Primary rocks and their mineralogical features. By contrast, Cuvier and his collaborator Brongniart concentrated on the rocks above the Chalk, Werner's Alluvial group. In an 'Essai sur la géographie mineralogique des environs de Paris' (1808) they showed that Werner's Alluvial rocks were a substantial and complex succession of Tertiary formations.

Cuvier and Brongniart found that strata of limestone, marlstone, and shale, even very thin ones, occurred in a predictable, fixed order of superposition throughout an extensive part of the Paris Basin: 'Cette constance dans l'ordre de superposition des couches les plus minces, et sur une étendue de 12 myriamètres au moins, est, selon nous, un des faits les plus remarquables que nous ayons constatés dans la suite de nos recherches.' They also found that the most reliable feature by which a bed can be characterized and its distant exposures be correlated is not its mineralogical composition, but its fossil content: 'Le moyen que nous avons employé pour reconnoitre au milieu d'un si grand nombre de lits calcaires, un lit déjà observé dans un canton très-éloigné, est pris de la nature des fossiles renfermés dans chaque couche, ces fossiles sont toujours généralement les mêmes dans les couches correspondantes, et présentent des différences d'espèces assez notables d'un système de couch à un autre système.'[10]

---

[7] See n. 1.

[8] Engelhardt, 'Die Entwicklung der geologischen Ideen seit der Goethe-Zeit', *Abh. Braunschweigischen Wissenschaftlichen Gesellschaft,* 1979, 1–23; Oldroyd, 'Historicism and the Rise of Historical Geology', *Hist. Sci.* xvii (1979), 191–213.

[9] 'Historical Eloge of Abraham Gottlob Werner', *Edinburgh Phil. Journ.* iv (1821), 2.

[10] 'This constancy in the order of superposition of the very thinnest beds over a distance of at least 120 kilometres is, we believe, one of the most remarkable facts which we have established in the course of our researches.'

In the Tertiary succession of the Paris Basin some beds contained marine shells, and others, alternating with these, the remains of terrestrial vertebrates. From this Cuvier concluded that the beds were the deposits of discrete periods of geological history, separated by an interchange of sea and land. A formation was thus not primarily defined by its composition, but by its duration; a formation represented the deposits of a particular period of earth history, best characterized by the remains of its faunal inhabitants. Cuvier wrote: 'It is to them alone that we owe the commencement of even a Theory of the Earth; as, but for them, we could never have even suspected that there had existed any successive epochs in the formation of our earth . . .'[11] This was an exaggeration, but Cuvier, more than Werner, directed the attention of contemporaries to the importance of fossils.

The early *Transactions of the Geological Society* show a gradual shift from Wernerian stratigraphy (focused on Primary rocks and on mineralogy) to Cuvierian historical geology (focused on fossiliferous rocks and on paleontology). Representative of the former were the contributions by John MacCulloch. The first volume of *Transactions* (1811) contained only one paper in which due attention was given to fossils. It was by James Parkinson on the strata of the London Basin.[12] This was followed by a study of the Hampshire Basin by Thomas Webster in the second volume of the *Transactions* (1814).[13] Both papers exemplified Cuvierian stratigraphy; moreover, they demonstrated a similarity of superposition of the Tertiary beds in England with those in France. They thus verified over a long international distance that formations occur in a fixed order of succession.

Parkinson not only drew attention to Cuvier and Brongniart, but also to William Smith. In addition to the Wernerians and to Cuvier and his school, Smith represented a third component in the fabric of

---

'The means which we have used to recognize a bed already observed in a very distant district amidst so large a number of limestone beds comes from the nature of the fossils enclosed in each bed; these fossils are always roughly the same in corresponding beds, and show a very noticeable difference of species from one group of beds to another.' *Annales du Muséum d'Histoire Naturelle* xi (1808), 307–8.

[11] *Theory of the Earth*, 1813, 54.

[12] 'Observations on Some of the Strata in the Neighbourhood of London', *Trans. Geol. Soc.* i (1811), 324–54.

[13] 'On the Freshwater Formations in the Isle of Wight', ibid., ii (1814), 161–254.

early nineteenth-century stratigraphy, namely the English provincial tradition. Smith and his circle initiated the English study of Secondary formations. As early as 1799 he had completed a list of the succession of strata in the vicinity of Bath. In the process he had discovered that English formations occur in a fixed order and that certain fossils are peculiar to certain strata. As late as 1815 he published his results, a map entitled *Delineation of the Strata of England and Wales, with Part of Scotland*, accompanied by a brief *Memoir to the Map*. In quick succession a number of other publications followed.[14] The nature of Smith's work was highly empirical and he rarely ventured a theoretical comment; but in his *Stratigraphical System* (1817) he did suggest that fossils represent the creations of past periods of earth history.

After the third volume of *Transactions* (1816) of the Geological Society had appeared, with further contributions to stratigraphy, Fitton reviewed the state of the subject for the *Edinburgh Review* (1817). He acknowledged Werner as 'having been the first to draw the attention of geologists, explicitly, to the *order of succession* which the various natural families of rocks are found in general to present, and in having himself developed that order to a certain extent, and with a degree of accuracy which, before his time, was unattainable, from the want of sufficient methods of discriminating minerals and their compounds'.[15] He also credited Werner with the concept of formations, and with the principle that the crust of the earth is composed of a series of these formations, laid over each other in a fixed order. Fitton explained this as follows: 'Thus, in the series A, B, C, D, it may happen that B or C, or both, may be occasionally wanting, and consequently D be found immediately above A; but the succession is *never* violated, nor the order inverted, by the discovery of A above the formations B, or C, or D, nor of B above those that follow it etc.'[16]

Fitton made no mention of Cuvier's contributions, and as such he showed his Wernerian colours. But in a second essay for the *Edinburgh Review* (1818)[17] Fitton did broach the delicate question of the relative merit of Smith's contributions to stratigraphy. Even

[14] See Sheppard, 'William Smith: his Maps and Memoirs', *Proc. Yorkshire Geol. Soc.* xix (1917), 73–253.

[15] 'Transactions', xxix (1818), 71.

[16] Ibid. In fact, the concept of formations had previously occurred in Georg Christian Füchsel's *Historia Terrae et Maris*, 1761.

[17] 310–37.

though he had been late to publish his results Smith had, early on, freely communicated these to his friends, e.g. to the two clerics Benjamin Richardson and Joseph Townsend. The latter had used some of Smith's work in his *Character of Moses Established for Veracity as an Historian* (1813). Fitton credited Smith with the independent discovery of the elements of stratigraphy and with an intangible influence on other English geologists.

However, this acknowledgement was not enough for Smith's supporters, who seemed to feel that the metropolitan establishment of the Geological Society was not prepared to give provincial effort the credit that was its due. This suspicion was fuelled partly by the fact that a map of England and Wales, rival to Smith's, was being prepared under the aegis of the Geological Society. This project was directed by Greenough during the 1810s with the assistance of some members of the society. Much of Smith's work was duplicated and, when Greenough's map was published in 1820, questions of its direct or indirect debt to Smith's map became a source of claim and counter-claim. John Farey, another mineral surveyor, presented the case for Smith in several letters to the *Philosophical Magazine*. He listed no fewer than sixteen 'claims to merit and originality' of Smith, one of which was that organic remains are not randomly distributed through the strata, but 'that some one or an assemblage of two or more of *these species of fossil Shells,* etc. may serve as new and *more distinctive marks, of the identity of most of the Strata* in England, than were previously known or resorted to, by mineralogists or others'.[18]

INTERNATIONAL CORRELATION

These maps, however, were not central to the making of an international standard of relative chronology, the stratigraphic column. By the middle of the 1810s little more existed than the results of local or national studies; even Werner's table, in spite of its aspirations to universality, was based on little more than the study of one region. Their integration into a single, international table required the definition of broad categories of rock-time based on analogous successions in different regions. This activity, stratigraphic classification, involved some abstraction from the minutiae of local detail. Neither Greenough nor Smith were concerned with this; their formations

---

[18]li (1818), 177. See also Farey, 'Free Remarks on Mr. Greenough's Geological Map', ibid., lv (1820), 379–83.

were as close as they could make them to actual particulars, not grouped into higher categories for correlation with foreign geology. To Greenough the geological map of England and Wales represented a unique mosaic of rock formations without analogy abroad. He, and some other members of the Geological Society, explicitly denied that an international standard stratigraphy was possible.

Greenough was one of those early nineteenth-century English geologists who objected to premature theorizing and advocated an empirical, inductive approach to the subject. With him as its first president, the Geological Society had adopted this approach as its official philosophy, symbolized by the quotation from Bacon's *Novum Organum* on the title-pages of its *Transaction*.[19] Members of the Society were, in the words of Fitton, 'neither . . . Wernerians nor Huttonians, but plain men'.[20] They cultivated the collection of data and the method of field work. The element of abstraction in stratigraphic classification was not congenial to this philosophy.

Kidd argued in his *Geological Essay* (1815) that Werner's table of formations was unreal; he did not believe in a fixed succession of formations and concluded that 'scarcely any point of general uniformity is observable; each group, and almost every distinct part of each group, having its own pecularities'.[21] Greenough argued even more strongly than Kidd against a standard stratigraphy. His *First Principles of Geology* (1818) was a remarkable catalogue of doubts and uncertainties illustrated by complexities and exceptions. Greenough ridiculed Wernerian theory: 'Virgil's peasant fondly imagined his native village a miniature of imperial Rome. With a corresponding love of generalisation, Werner imagined Saxony and Bohemia a miniature of the world'.[22] Greenough rejected all criteria by which the relative age of rocks might be determined, including the criterion of fossil content.

It is ironic that Buckland, Kidd's successor at Oxford, who acknowledged Greenough as his mentor, took the initiative in producing an international standard stratigraphy. Younger

---

[18]'Quod si cui mortalium cordi et curae sit, non tantum inventis haerere, atqui iis uti, sed ad alteriora penetrare: atque non disputando adversarium, sed opere naturam vincere; denique non belle et probabiliter opinari, sed certo et ostensive scire; tales, tanquam veri scientiarum filii, nobis (si videbitur) se adjungant'.

[20]*Edinburgh Review* xxix (1818), 70.

[21]149.

[22]234. See Laudan, 'Ideas and Organisation in British Geology', *Isis* lxviii (1977), 527–38.

members of the Geological Society, such as Buckland who joined in 1813, paid lip-service to its Baconian ideals, but they readily committed themselves to theory. Buckland's philosophy can best be characterized as eclectic, neither Wernerian nor Huttonian, but considering itself free to choose and combine whatever elements of either theory appeared applicable. From the Wernerians Buckland borrowed their stratigraphic system, and from the Huttonians their theory of denudation, which attributed a large proportion of rocks to a detrital rather than a chemical origin. For his lectures Buckland used Playfair's illustrations of Hutton's *Theory of the Earth.* In 1817 he wrote to Greenough:

In Bath I attended two lectures of a young disciple of Jameson, Dr Galby who is very promising. He gave us Jameson's last theory of the coal formation, which is that all the substance of coal as well as the strata attending it is of chemical origin and unconnected with vegetable matter and that whin dykes are contemporaneous concretions in the rocks they traverse. The whole theory of denudation is to him a terra incognita. All pebbles in every species of conglomerate are considered as contemporaneous concretions, with sundry other monstrous opinions which at this time of day astonish me in Jameson.[23]

What Buckland did share with Greenough and his generation was reliance on field work on a large, even international scale. Frank Buckland has told how his father in the summer of 1808 began to explore the English countryside. This was his first geological excursion, and it was followed by others in 1809 and 1810. In 1811 he included Scotland and Ireland in his itinerary. In 1812 Buckland began to collaborate with J. J. and W. D. Conybeare; and in the course of 1812–15 he joined Greenough on extensive tours to collect data for the preparation of the geological map of England and Wales. From 1812 to 1824 Buckland rode once or twice a year from Oxford to Axminster visiting places along the way such as Bath, where he met Richardson and Townsend.[24] It was not until the winter of 1819–20 that Buckland and Smith met.

The end of the Napoleonic wars stimulated international travel and Buckland visited continental Europe several times. A particularly fruitful co-operation existed for some time between Buckland,

[23] Dated only by year, CUL, Gr P.
[24] F. T. Buckland, 'Memoir of Buckland', *Geology and Mineralogy,* i (3rd edn., 1858), xxvii–xxix.

Conybeare, and Greenough. Early in 1816 Buckland wrote to the latter:

I have received from W. Conybeare a most important communication of which the object is to establish between you, him and myself a geological triumvirate which in the course of the next summer shall spread conquests more extensive over the subterraneous world than were ever accomplished by our less penetrating predecessors, the superficial triumvirs of Rome. I am so thoroughly convinced that by working thus in concert we should do more in three months together, than singly in three years that I am disposed to make almost any sacrifice for the accomplishment of so important an union. I have all but absolutely engaged to be one of the party on condition that you can manage to make the third, and should this triple alliance be consummated would make every effort to be ready to move the beginning of June.[25]

That summer of 1816 Buckland went on an extensive tour of continental Europe. In the company of colleagues he examined foreign geology, visited grand old men like Goethe and Werner, exchanged ideas, collected maps, books, and specimens, and wrote glowing accounts of his experiences to friends back home.

Buckland also relied on work done by others. He used such interesting items as a hand-written list of the succession of formations in Saxony given to him by Werner; or a list of the succession in the vicinity of Bath obtained from a member of Smith's provincial circle.[26] Various people were directed by Buckland to explore distant stratigraphy. Among them was one of his students, William Strangways, who was the first person to have been examined at Oxford in the subject of geology. This happened in 1816 in the presence of von Buch who was visiting Buckland at the time. The examination questions were essentially problems of stratigraphy posed as questions in logic. One of these asked: 'Chalk is above Mountain Lime. Mountain Lime is above Granite. Chalk is above Granite. Is this a good syllogism?' Strangways's answer was true to his education: 'Chalk is above Mountain Lime etc.—a very good geological syllogism.'[27] Strangways subsequently left for Russia; his stay there added up to an exemplary story of foreign study. From Russia he sent back to Buckland and to the Geological Society boxes with specimens, geological maps, stratigraphic sections, and

[25] 21 Feb. 1816, CUL, Gr P.
[26] Miscellaneous stratigraphy file, OUM, Bu P.
[27] Ibid. See also Buckland to Mary Cole, 3 June 1816, NMW, Bu C.

honorary memberships for Buckland and others of the Imperial Societies of Mineralogy and Natural History at St. Petersburg and Moscow. Informative papers by Strangways in the *Transactions of the Geological Society* provide a lasting record of his Russian expedition.[28]

Another of Buckland's continental correspondents was the Austrian Count Breunner. In 1819, on Buckland's recommendation, he was given an honorary degree in civil law at Oxford. Breunner accompanied Buckland on more than one geological tour; and on his return to Austria provided Buckland with data on the stratigraphy of the Vienna Basin. One of Breunner's letters informed Buckland that 'the balls and gayeties of the town took the greatest part of my time but very soon after when the weather was fine I tore myself from those gayeties, to the great astonishment of the whole Vienna world, particularly of the young ladies who thought it impossible that I could resist their charms.'[29]

Buckland's knowledge of stratigraphy was not restricted to Europe; data obtained from civil servants in the overseas colonies extended his information to cover other continents. This was accomplished with the help of his governmental Tory friends. In particular Bathurst, who was secretary for the colonies, the chancellor of the university, and the foreign secretary were instrumental in adding to Buckland's store of colonial geology. He wrote to Mary Cole in 1819:

You will be pleased to hear I am likely to get extensive importations from all the British Colonies over the world through the kindness of Lord Bathurst who lately sent me a message requesting I would draw up a list of instructions for collecting specimens in geology of which he would transmit copies to all the colonies connected with his office and adding that it is his intention to deposit the specimens that may be sent home in the Oxford Museum for the purpose of illustrating my lectures. Lord Grenville is much pleased at this very handsome proposal on the part of the noble earl and has written to Lord Castlereagh requesting him to transmit my papers to the King's Ministers abroad connected with his office and enlist them also in the noble cause of geology, which by the creation of a new professorship and other events that have occurred since I last saw you has I trust been permanently established amongst the sciences that are taught in the University.[30]

[28] E.g. 'Geological Sketch of the Environs of Petersburg', *Trans. Geol. Soc.* v (1821), 392–458.
[29] 6 Apr. 1821, OUM, Bu P.
[30] 29 Oct. 1819, NMW, Bu P. See also 5 Jan. 1820, ibid.

Some collections did arrive from overseas, e.g. one from Madagascar, and another from New South Wales. In a paper on these presented to the Geological Society (1821) Buckland commented that the collections were 'a valuable addition to that stock of information which is gradually advancing the science of geology towards a point, at which it will be competent for us to form some well grounded theory as to the general structure of the earth'.[31] Examination of the specimens encouraged Buckland in the belief, contrary to that expressed by Greenough and Kidd, that a standard stratigraphy, applicable to the whole world, was possible. 'It is satisfactory to find,' Buckland concluded his brief paper, 'on comparing rocks from such remote parts of the southern hemisphere with those of Europe, that none of them afford any varieties that may not be referred to species that occur also on this side of the equator, and that as far as they go, they lead us towards a conclusion, that there is not only an identity in the older formations of rocks that constitute the earth's surface, but also a strong resemblance in the leading features of many of the secondary strata that follow and repose upon them.'[32]

This conviction helped Buckland put together a stratigraphic column for the British Isles and a table comparing the British succession with that of continental Europe. The British column took shape by way of a number of successive sheets prepared and printed for the illustration of Buckland's geology lectures. Four different editions exist (1814, 1815, 1816, 1818).[33]

The changes and additions to successive versions record the rapid advances of stratigraphy during the 1810s. There is a noticeable shift away from the Wernerian preoccupation with Primary rocks towards studies of Secondary and Tertiary successions. The first two editions were in part based on continental examples; the third and fourth no longer needed these and used almost entirely British localities and were entitled: 'Order of Superposition of Strata in the British Islands'.

The table of 1818 was a milestone in the progress of British stratigraphy. It brought together a large number of local observations; a

[31] 'Notice on the Geological Structure of a Part of the Island of Madagascar', *Trans. Geol. Soc.* v (1821), 476.

[32] Ibid., 481. Buckland's negative view of Greenough's *First Principles* was expressed in a letter to Mary Cole, 18 May 1820: 'I have a larger class than ever in spite of all that Mr. Greenough has done in his book to discourage geology.' NMW, Bu P.

[33] Miscellaneous stratigraphy file, OUM, Bu P.

substantial portion of the succession was very nearly complete; and the multiplicity of actual strata was grouped into higher categories of classification for the purpose of international comparison and correlation. In particular the strata known later as the Jurassic were nearly complete, and the names and examples showed Buckland's expertise in the Oxford area, and his use of Smith's table of the vicinity of Bath.[34] The strata were listed by name, and to each name was added type localities, main characteristics, and maximum observed thickness. They were grouped into ten higher, formational units, and these were related to the fourfold Wernerian division of earth history.

The 1816 edition, the first with the title 'Order of Superposition of Strata in the British Islands', was widely distributed. Copies were handed out at Oxford, in London, and sent to colleagues at home and abroad. The table represented a major step towards a single European classification of formations. One of the continentals who received a copy was von Buch; he praised Buckland highly for his pioneering effort, and in a letter of 1817 he wrote, in characteristically Germanic English:

Your classification of English strata is very important. It gives my great desire to make it known in Germany with some remarks which I have learnt in your and your friends conversations. Having such a plan in hand, we may easier compare the different countrys. It is the map of the Geology. I hope you will not loose courage to bring the two first class to such a perfection as the secondary is. It is certainly possible with attention and a sufficient mass of observations; and a great instigation to do it, is I think, the observation that in general, disorder is only found every where by idiots, even where the greatest order opens itself to the attentive and instructed naturalist.[35]

Another who, through Arthur Aiken at the Geological Society, received a copy of Buckland's table of 1816 was William Phillips. Phillips was at that time preparing a new edition of his *Outline of the Geology of England and Wales*. At his request Buckland updated his table and let it be printed in the new 1818 edition. Phillips commented: 'This classification is necessarily theoretical in part, and professes to shew the general agreement of the British strata, with the order adopted by the Wernerian School. It cannot fail to interest and instruct the student, as exhibiting at one grand view the great series

[34] See Arkell, *Jurassic System in Great Britain*, 1933, 4; see also ibid., 7.
[35] Dated only by year, RS, Bu P.

of our rocks, and as containing many localities which have not I believe been made public in any other form.'[36]

Specimens were also sent to Buckland from America. These further strengthened his belief in a similarity of rock types and their succession on different continents. In a letter in Silliman's *Journal* (1822) Buckland referred to the 'Geological Analogy recently discovered to exist between the two great Continents of the old and the new world'.[37] But specimens collected by others were not enough; during the summer of 1820 Buckland revisited continental Europe where he examined and drew up a number of sections of different rock successions, particulary in the Alpine region. On these he coloured in each particular formation, identified and delineated by its overall lithology and by physical breaks or discontinuities in the sequence. These he compared with their British counterparts. On the basis of such sections he constructed a comparative table of continental and English stratigraphy which he presented as a paper to the Geological Society. It was published in the *Annals of Philosophy* (1821) and a twenty-page offprint was made for distribution to colleagues at home and abroad.[38] Mantell reproduced part of it in his *Fossils of the South Downs* (1822), and among continental geologists Humboldt praised Buckland's table.

It contained a clear statement of his philosophy and methods of stratigraphy. Buckland wrote about previously mistaken identifications of continental rocks:

I shall hope, however, to prove their identity with English formations by the evidence of actual sections; and to show that a constant and regular order of succession prevails in the alpine and transalpine districts, and generally over the Continent, and that this order is the same that exists in our own country. But though referable to the same system, and coeval in point of time, and conformable with respect to their relative order of succession, the formations of England and the Alps are much disguised by local circumstances, and present widely varying features, the extremes of which it would be impossible to identify without the fortunate interposition of certain connecting links that are equally related to, and partake equally of, the characters of them both.[39]

[36] 19–20. See Phillips to Buckland, 18 Dec. 1817. OUM, Bu P.
[37] *Am. Journ. Sci.* iv (1822) 185–6.
[38] 'On the Structure of the Alps and Adjoining Parts of the Continent', *Ann. Phil.* i (1821), 450–68.
[39] 451.

The 1821 table of international correlation and classification was remarkably advanced. The only major deficiencies were the lack of a subdivision of the Transition series and of a sharp Permo-Triassic boundary; apart from this all major formations or systems of the definitive version of approximately 1840 were recognized: (1) Transition Limestone and Greywacke (Cambrian, Ordovician, and Silurian), (2) Old Red Sandstone (Devonian), (3) Carboniferous Limestone (Lower Carboniferous or Mississippian), (4) Coal Measures (Upper Carboniferous or Pennsylvanian), (5) New Red Conglomerate (Permian), (6) New Red Sandstone and Magnesium Limestone (Permo-Triassic), (7) Oolite or Jura Limestone (Jurassic), (8) Chalk and Greensand (Cretaceous), (9) Tertiary, (10) Diluvium, (11) Alluvium.

Buckland defined a formation, as had Cuvier, as the rock record of a particular period of earth history, 'a collected mass of beds, exhibiting identity of circumstances and time, not of materials'.[40] This last part of the definition was a proviso intended to allow for the minutiae of local detail and to dissociate Buckland from the Wernerian theory of a universal ocean. However, lithological features remained the prime criterion for correlation and classification, indicated by such formational names as 'Coal Measures' and 'Chalk'. The classification of strata into formational groups or systems was based on an assessment of the relative importance of changes of composition and, to a lesser extent, of degree of deformation. Buckland was well acquainted with the fossil discoveries of both Cuvier and Smith; but these had not yet added up to an independent criterion of relative age. Fossils were listed in the compositional description of rocks; trilobites, for example, were characteristic of certain Transition Greywacke slates on both sides of the Channel.

The circumstances under which the skeleton column of 1821 originated show that neither the theory of evolution nor that of uniformity had any part in the construction of the new, relative timetable of earth history.

## A PLETHORA OF TABLES

The year 1821 was a watershed in the development of stratigraphy. About a decade later MacCulloch stated that since then nothing

[40] Jackson, 'Buckland's Geological Lectures', 1, IGS, Bu P.

new had been added to the subject. This, of course, was a Wernerian view, because much of the paleontological enrichment of the stratigraphic column took place during the 1820s. In 1821 MacCulloch himself published *A Geological Classification of Rocks*, an explicit Wernerian mineralogical treatment of stratigraphy concentrated on the lower crystalline portion of the succession. A similar Wernerian classification was proposed by Thomas Weaver in the *Annals of Philosophy* (1821, 1822) who disputed some of Buckland's suggested correlations of English geology with Werner's table.[41] Yet another Wernerian, Aimé Boué, issued a very elaborate 'Synoptical Table of the Formations of the Crust of the Earth' in the *Edinburgh Philosophical Journal* (1825).[42] Boué's table was based on many more examples than Buckland's and corrected the position of the Muschelkalk.

Boué no longer gave preponderance to Primary rocks, as Wernerians had been inclined to do before him, but included the substantial additions of Secondary and Tertiary stratigraphy. Other Wernerians did the same, especially Humboldt, whose treatise on stratigraphy was published simultaneously in French and German in 1823 and translated that same year into English under the title *A Geognostical Essay on the Superposition of Rocks, in both Hemispheres*. Parts of it were serialized in the *Edinburgh Philosophical Journal*. Humboldt's world travels, in particular to Central and South America, gave his views on stratigraphic classification the factual basis which Werner's had so sorely lacked. He took issue with those who had argued 'that each portion of the globe differed in its geological constitution'. When international study began, 'The most striking analogies in the position, composition, and the included organic remains of contemporary beds, were then observed in both hemispheres; and,' Humboldt continued, 'in proportion as we consider *formations* under a more general point of view, their *identity* daily becomes more probable.'[43]

The 1820s were rich in tables of international correlation and classification. A prominent example from the French was Brongniart's elaborate *Tableau des terrains qui composent l'écorce du globe* (1829). Such attempts to unify the many stratigraphic studies from different

[41] iv (1822), 81–98.

[42] xiii (1825), 130–45.

[43] *Geognostical Essay*, 1823, 3. See also Baumgärtel, 'Alexander von Humboldt', *Toward a History of Geology*, ed. Schneer, 1969, 19–35.

countries and continents went hand in hand with proposals to standardize nomenclature. With characteristic French aplomb Brongniart introduced an entirely new set of names for all the major and for some of the lesser formational divisions. This produced feelings of rancour among those who, like Featherstonhaugh, were proud of the many English names in the stratigraphic vocabulary. He wrote to Buckland: 'The French philosophers are trying to eclipse the celebrity of the English geologists. They are as usual putting ruffles and corsets to the science. Make war against them in phalanx.'[44] And Featherstonhaugh followed his own advice when Amos Eaton proposed nationalistic innovations of classification and nomenclature for American stratigraphy, something Featherston-haugh's *Monthly American Journal* (1831–2) contemptuously discarded.[45]

However, English geologists had also been guilty of revisions of nomenclature. Conybeare, in his *Outlines,* replaced the Wernerian terms of Primary, Transition, Secondary, and Tertiary with new descriptions of his own. Using the Carboniferous as his point of reference, he proposed the terms Inferior, Submedial, Medial (Carboniferous), Supermedial, and Superior. A further revision was suggested by de la Beche (1827)[46] Fortunately, these changes did not catch on; Humboldt advised geologists to avoid 'national vanity' and keep to already-accepted names, especially in cases where formations were best known from particular areas such as the Lias in England and the Muschelkalk in Germany[47]—and this advice was generally followed.

Buckland's table of 1821, which owed much to the work done by Conybeare, became subsumed in the latter's *Outlines,* a further revised and enlarged edition of the book on the geology of England and Wales by Phillips. This latest edition was a classic of early English stratigraphy. Conybeare's 'Memoir' on the geology of Europe, however, published like Buckland's table in the *Annals of Philosophy* (1823), did not have the same lucid originality, but contributed to the alienation of Greenough from Buckland and Conybeare.[48]

[44] 27 June 1829, RS, Bu C, fo. 32.
[45] i. 82–91.
[46] *A Tabular and Proportional View . . . ,* 1827.
[47] 'On Rock Formations', *Edinburgh Phil. Journ.* x (1824), 234.
[48] v (1823), 1–16, 135–49, 210–18, 278–89, 356–9; vi (1823), 214–19. See Buckland to Greenough, 22 June 1823, CUL, Gr P.

During the 1820s Buckland remained at the centre of English stratigraphy; new data were continuously sent to him from such varied sources as Mary Cole in the west, the astronomer G. B. Airy in the east, and the botanist N. J. Winch in the north.[49] New observations also reached him from the Continent, for example by Joseph Pentland who examined the rock succession in the Apennines, particularly in the vicinity of Rome.[50] Buckland contributed two further major studies of English stratigraphy which were model studies of regional geology.[51]

These local studies were added to the successively improved and enlarged editions of the international standard column of formational succession and firmly established the fundamentals of stratigraphy. By about 1830 the fact of a standard column was no longer a subject of disagreement. A fairly uniform stratigraphic table had become widely accepted; such tables were no longer connected with any particular author, but had become common textbook property, part of the alphabet of geology. A popular periodical like Loudon's *Magazine of Natural History* (1830) introduced its readers to the 'geological systems of arrangement' and to the 'order of creation' which the successive formations represented.[52] In review papers the subject of stratigraphy and its relative timetable of earth history were treated as facts of geology. Wall charts appeared which confidently and in some detail depicted the crust of the earth in cross section, composed of formations in a fixed and universal succession. Nérée Boubée's chart (1833) was probably the most widely used example (figure 11).

The additions to stratigraphy of the 1830s were of a different kind from the work of the 1820s; the subject entered a more professional phase during which the focus of interest was not so much on overall outline, but on specifics of boundary definition and statistical use of fossils to define and refine single systems and their subdivisions. The 1830s were the decade of Lyell's subdivision of the Tertiary (1833), of the definition of the Cambrian by Sedgwick (1835), of the Silurian (1835) and Permian (1841) by Murchison, and of the

---

[49] E.g. Airy to Buckland, 15 July 1830, DRO, Bu P.; Buckland to Winch, 2 Apr. 1821, LS, Wi P.

[50] Miscellaneous stratigraphy file, OUM, Bu P.

[51] In collaboration with Conybeare, 'Observations on the South-Western Coal District of England', *Trans. Geol. Soc.* i (1824), 210–316; jointly with de la Beche, 'On the Geology of the Neighbourhood of Weymouth', ibid., iv (1836), 1–46.

[52] iii (1830), 62–78.

Devonian by Murchison and Sedgwick (1839). It was the decade of the redefinition of the old Wernerian categories by Sedgwick (1838) and Phillips (1841) and their replacement by the now familiar 'Paleozoic', 'Mesozoic', and 'Cenozoic', based on the stratigraphic distribution of fossil types.[53] Buckland's involvement in this work was mostly indirect; he suggested to Murchison the locality for his Silurian work, included the Silurian in his Bridgewater Treatise,[54] and acted as mediator during the fierce debates which boundary definitions generated.

[53] Phillips, *Figures and Descriptions of the Palaeozoic Fossils*, 1841, 160.

[54] Murchison announced his Silurian system in a letter to Buckland, 17 June 1835, OUM, Bu P.

# 11
# Inhabitants of Former Worlds

## VERTEBRATE EXTINCTIONS

The Wernerian study of rock types relied on auxiliary subjects such as chemistry and mineralogy. The systematic study of fossils, however, had to be based on the life sciences. Expertise in comparative anatomy was acquired by several English geologists during the 1810s, and in the course of the following decade interest in vertebrate fossils mushroomed. This change was to a large extent inspired by Cuvier's comprehensive study of fossil quadrupeds, by Buckland's cave paleontology, and by the discovery of a number of sensationally unfamiliar reptilian fossils.

During the 1820s the skeleton of stratigraphic succession acquired the flesh of paleontology. By about 1830 the study of fossils and of fossiliferous strata dominated English geology. In the process 'mineralogical geology' and 'zoological geology' were fused to become historical geology in its modern form. The nature of the fossils, seen in the context of their stratigraphical sequence, revealed a historical panorama of major changes in climate, in the distribution of land and sea, and in the character of life itself. The earth appeared to have passed through a succession of former worlds, of periods of history during which our planet was populated by seemingly alien inhabitants quite different from those of today and now long since extinct. Thus historical geology demonstrated that the early modern notion of a plurality of worlds was correct, if not in space, at least in time.

This trend from mineralogy to paleontology is apparent in the papers read to the Geological Society. The isolated contributions by Parkinson (1811) and Webster (1814) to 'zoological geology', then still unfashionable, were followed during the 1820s by a variety of classic papers on paleontology by Buckland, Conybeare, de la Beche, Mantell, and others. Major studies, especially of vertebrate paleontology, also appeared in the *Transactions* of the Royal Society.

The *Annals of Philosophy* served as the medium for rapid notification of the results of English paleontology.

These studies of fossils from Secondary and Tertiary rocks displaced the Wernerian interest in Primary rocks and their mineralogical properties. Predictably Jameson persisted for some years with an editorial policy not to report on paleontology in his *Edinburgh Philosophical Journal*, but by the middle of the 1820s he joined Wernerians who had yielded to the novel interest. Fitton, however, deplored the drift away from mineralogy in his anniversary address to the Geological Society (1829):

Mineralogy has, from various causes, been of late less vigorously pursued in England, than a few years ago . . . The naturalist, however, who is in search of general laws, should exert himself to keep every part of his subject in view; and should never cease to remember, that, as in the study of the newer formations Zoology and Botany are his best allies, —so Mineralogy is indispensable to an acquaintance with the more ancient rocks, —and Chemistry as well as general Physics, to the solution of the problems connected with them.[1]

Fitton's presidential lament was an indication of the remarkable change which the subject of geology had undergone by the end of the 1820s. Cuvier, more than anyone else, was acknowledged as the scientist who brough zoology to bear on geology. The first two editions of his monumental and voluminous *Recherches sur les ossemens fossiles* (1812; 1821–4) were the main inspiration and rallying point for the first generation of nineteenth-century English paleontologists. Cuvier was a great comparative anatomist who applied his skill systematically to the study of fossil tetrapods or quadrupeds, i.e. amphibians, reptiles, birds, and mammals.

Fossil quadrupeds rarely consist of perfectly preserved specimens, but mostly of a few loose bones and teeth. With the aid of comparative anatomy it becomes possible to identify the kind of animal and to reconstruct the entire skeleton from its few incomplete remains. The latter was achieved on the basis of the 'law of organic correspondence', formulated by Cuvier as follows: 'Every organized individual forms an entire system of its own, all the parts of which mutually correspond, and concur to produce a certain definite purpose, by reciprocal reaction, or by combining towards the same end. Hence none of these separate parts can change their forms

[1] *Proc. Geol. Soc.* i (1834), 121.

without a corresponding change on the other parts of the same animal, and consequently each of these parts taken separately, indicates all the other parts to which it has belonged.' 'Thus,' Cuvier boasted rather overconfidently, 'commencing our investigation by a careful survey of any one bone by itself, a person who is sufficiently master of the laws of organic structure, may, as it were, reconstruct the whole animal to which that bone had belonged.'[2]

The identification of many quadrupedal remains established the fact of extinction against all expressed doubts. It became abundantly clear that not only species but even genera and higher taxonomic units had been annihilated.[3] Without extinction it was difficult to explain the huge mammalian remains discovered mainly in North and South America. Bones were unearthed of such gigantic fossil beasts as the mastodon and mammoth (members of the elephant family) and of the megalonix and megatherium (allied to the sloth). Those who opposed the possibility of extinction believed that the unfamiliar size and shape of the fossils had been produced by environmental influences such as climate; or they suggested that the mammoth, for example, was a fabulous monster, just like the centaur; others speculated that the mammoth might still inhabit unexplored regions.[4]

But the reality of the mammoth was vividly illustrated in 1799 by the discovery of a complete specimen, its flesh and hair intact, preserved in the Siberian permafrost. It was brought to international attention in 1806 by a member of the Imperial Academy of Sciences of St. Petersburg, and small specimens of mammoth hair were despatched to a number of scientific centres.[5] As for the hypothesis that very large mammals like the mammoth or the mastodon might still be alive, Cuvier countered that 'it is quite impossible to conceive that the enormous *mastodontes* and gigantic *megatheria,* whose bones have been discovered in North and South America, can still exist alive in that quarter of the world. They could not fail to be observed by the hunting tribes, which continually wander in all directions through the wilds of America.'[6] In his *Ossemens fossiles* Cuvier presented a systematic description of these mammalian

[2] *Theory of the Earth,* 90, 95.
[3] See ch. 14.
[4] See White, 'Observations on a Thigh Bone of Uncommon Length', *Memoirs Lit. Phil. Soc. Manchester* ii (1785), 350–7. See also Ashe, *Memoirs of Mammoth,* 1806, 49.
[5] Anon., *On the Mammoth or Fossil Elephant,* 1819, 14–15.
[6] *Theory of the Earth,* 86.

remains in the superficial deposits of the Diluvium. From older, Tertiary rocks Cuvier added further examples of extinct mammals, e.g. the paleotherium or the anoplotherium, fossil pachyderms found in the gypsum quarries of Montmartre near Paris. The collective force of these and other examples sufficed to establish the fact of extinction to the satisfaction of most contemporary naturalists. A paleontologist was to some extent as good as his fossil collection. Civic or even national pride could be based on the possession of particular specimens or of collections, especially those of large vertebrate fossils. Private, provincial, and metropolitan museums at times vied with each other to acquire rare and fine specimens. Some of these could fetch very substantial prices; a rare reptilian fossil might sell for around £150 and private collections for several thousand pounds.[7] Fossil collecting could become a livelihood, as in the case of the famous collector from Lyme Regis, Mary Anning, to whom early paleontology was indebted for the discovery of a variety of Liassic reptiles.

During the 1820s Buckland avidly added to his private collection from as many sources as possible. Soon his rooms in Corpus Christi College became a legendary jumble of objects of natural history. In 1821 Duncan was inspired to versify a 'Picture of the Comforts of a Professor's Rooms in C.C.C., Oxford':

> Here see the wrecks of beasts and fishes,
> With broken saucers, cups, and dishes;
> The pre-Adamic systems jumbled,
> With sublapsarian breccia tumbled,
> and post-Noachian bears and flounders,
> With heads of crocodiles and founders;
> Skins wanting bones, bones wanting skins,
> And various blocks to break your shins.
> No place in this for cutting capers,
> Midst jumbled stones, and books, and papers,
> Stuffed birds, portfolios, packing-cases,
> And founders fallen upon their faces.[8]

That year saw the addition to Buckland's collection of vertebrate fossils from Kirkdale Cave; these formed the core of his growing geological museum. Collecting fever gripped him when de la Beche

---

[7] Mantell, 'A few Notes on the Prices of Fossils', *London Geol. Journ.* i (1846), 13–17.
[8] Daubeny, *Fugitive Poems*, 1869, 81–2.

sent him a rare reptilian jaw bone: 'I have been more delighted', Buckland exclaimed, 'than with any thing fossil I ever yet beheld by the sight of your glorious jaw of Plesiosaurus which deserves to be cast in gold and circulated over the universe.'[9] He called his own geological collection 'decidedly the richest and most instructive of its kind in the world'.[10] His museum continued to expand during the 1820s and 1830s, a goal in pursuit of which he cultivated the acquaintance of Arctic explorers, in particular of naval captains such as Beechey, Parry, or Smyth, whom he invited to such ceremonial occasions as the university's Encaenia.[11] Specimens were regularly donated by people as varied as Grenville, the chancellor of the university, and Strangways, one of Buckland's early students.[12]

THE AGE OF THE REPTILES

The fossils which during these early days of modern paleontology shaped the popular imagination of extinct life and former worlds were two aquatic reptilian forms, ichthyosaur and plesiosaur, and two terrestrial ones, megalosaur and iguanodon. Their discovery occurred within the decade 1814 to 1824. The type specimens all came from localities in the Secondary formations of England, the aquatic reptiles from the Lias between Lyme Regis and Charmouth on the Dorset coast, where Mary Anning collected the main specimens. The ichthyosaur ('fish lizard') was described by Home in a series of papers, the first in 1814,[13] and was named in 1818 by König. The classic study of the ichthyosaur, however, in which the plesiosaur ('akin to the lizard') was also described, was by Conybeare assisted by de la Beche (1821).[14] Further well-preserved specimens were discovered, as a result of which both the ichthyosaur

[9] Buckland to de la Beche, 20 Apr. 1822, NMW, Bu P.

[10] 9 Sept. 1823, ibid.

[11] E.g. Parry to Buckland, 6 Apr. 1829, 13 June 1829, and Smyth to Buckland, 21 Mar. 1833; DRO, Bu P. See also Gordon, *Life and Correspondence of Buckland,* 1894, 44–50.

[12] See Buckland to Warburton, 4 Dec. 1820, CUL, Wo P.

[13] 'Some Account of the Fossil Remains of an Animal more nearly Allied to Fishes than to any of the other Classes of Animals', *Phil. Trans.* civ (1814), 571–7. See also Delair, 'A History of the Early Discoveries of Liassic Ichthyosaurs', *Proc. Dorset Nat. Hist. Archaeol. Soc.* xc (1968), 115–27.

[14] 'Notice of the Discovery of a New Fossil Animal, Forming a Link between the Ichthyosaurus and the Crocodile', *Trans. Geol. Soc.* v (1821), 559–94. On Pentland's contribution to this paper see Delair and Sarjeant, 'Joseph Pentland', *Proc. Dorset Nat. Hist. Archaeol. Soc.* xcvi (1975), 12–16.

and the plesiosaur could be reconstructed in detail and with considerable accuracy. The former could reach a length of up to thirty-five feet; the latter was smaller, though its appearance more unusual. Far less complete were the remains from which the two terrestrial reptiles were identified. Bones and teeth of the megalosaur ('great lizard') were apparently in Buckland's possession as early as 1818; but he published his discovery several years later (1824).[15] The iguanodon was described by Mantell the following year, although as early as 1822 he possessed teeth of this ancient reptilian monster.[16] Mantell noticed the similarity between the fossil teeth and those of the living Iguana; hence the name iguanodon ('teeth like those of the Iguana').

Initial estimates of the size of the terrestrial reptiles were exagerated. Buckland conjectured that the megalosaur had been as long as sixty to seventy feet; and Mantell calculated that the iguanodon reached up to 100 feet from head to tail. Owen later reduced these numbers considerably, using novel assumptions for anatomical comparisons. Nevertheless, these ancient land reptiles reached prodigious dimensions, and because of this Owen coined the collective name of 'dinosaurs' ('terribly great lizards'). Unfortunately, because complete fossil skeletons were lacking, early attempts to reconstruct the dinosaurs were far from accurate. Thus they were depicted as standing on their four legs; in reality, both the megalosaur and the iguanodon were largely bipedal and walked on their hind legs.

The first decade of reptilian paleontology came to an end with the completion of the second edition of Cuvier's *Ossemens fossiles*, which contained much of the new material. The reptilian fossils were an impressive contribution by the English. De la Beche sketched a coat of arms for the Geological Society which was flanked by an ichthyosaur on the left and a plesiosaur on the right.[17] When Conybeare reviewed the state of geology in his report to the British Association, he listed as the main paleontological achievement of the 1820s the discovery of the ancient reptiles; 'this period', he said, 'has witnessed the complete restitution, and I may almost say the resurrection, of the long-extinct and monstrous Saurians of the lias; the

---

[15] 'Notice on the Megalosaurus', *Trans. Geol. Soc.* i (1824), 390–6.

[16] 'Notice on the Iguanodon', *Phil. Trans.* cxv (1825), 179–86. See also Delair and Sarjeant, 'The Earliest Discoveries of Dinosaurs', *Isis* lxvi (1975), 5–25.

[17] De la Beche to Buckland, 13 June 1823, DRO, Bu P.

oolites of Stonesfield and the Wealden limestone of Tilgate have yielded the Megalosaurus, and the Iguanodon, to the researches of Buckland, and Mantell'.[18]

New discoveries continued to be made, and were described in articles and books, among which was Thomas Hawkins's *Memoirs of Ichthyosauri and Plesiosauri* (1834), dedicated to Buckland and Conybeare. Owen's reports on British fossil reptiles to the meetings of the British Association of 1839 and 1841 constituted a landmark in the classification of fossil reptiles. Owen listed no fewer than ten species of ichthyosaurs and sixteen species of plesiosaurs.[19]

Some mammalian fossils, though large in size, were not unfamiliar in shape. The mammoth, for example, looked much like living elephants. The reptilian fossils, however, which were derived from deposits very much older than those from which the mammalian remains had been recovered, were not only large in comparison to living reptiles, but unfamiliar in form. The skeletons of the aquatic reptiles in particular seemed composed of disparate anatomical elements. They were reminiscent of the dragons of myth and legend. The plesiosaur with its serpentine neck looked especially like a conventional dragon. Contemporary writers delighted in descriptions of the bizarre appearance of these monsters. There could not have been a more dramatic demonstration of the discovery of alien life forms from a former world. Witness Buckland's own description of the anatomy of the ichthyosaur: 'Thus, in the same individual, the snout of a Porpoise is combined with the teeth of a Crocodile, the head of a Lizard with the vertebrae of a Fish, and the sternum of an Ornithorhynchus with the paddles of a Whale.'[20] About the plesiosaur he wrote: 'To the head of a Lizard, it united the teeth of a Crocodile; a neck of enormous length, resembling the body of a Serpent: a trunk and tail having the proportions of an ordinary quadruped, the ribs of a Chameleon, and the paddles of a Whale.'[21]

The behaviour of these reptiles was imagined as dragon-like. Pidgeon conjectured that they 'were evidently fierce and rapacious reptiles,' and that 'the ichthyosaurus must have been an overmatch for its antagonist, unless the flexible neck of the latter gave it some

[18] *Report BAAS, 1831, 1832,* 402.
[19] *Report BAAS, 1839,* 13–126; *1841,* 60–204.
[20] *Geology and Mineralogy,* i (1836), 169.
[21] Ibid., 202–3.

advantages on the score of activity'.[22] Even though the iguanodon was a herbivore it was habitually depicted as locked in mortal combat with the megalosaur, struggling in the mud of ancient lowlands. From children's books such as Hack's *Geological Sketches, and Glimpses of the Ancient Earth* (1832) to Hawkins's idiosyncratic *Book of the Great Sea Dragons* (1840) literary imagination went far beyond the facts of paleontology in casting the fossil reptiles in the role of fabled dragons.

Even though most excitement was engendered by the discovery of vertebrate fossils, significant advances were made in the field of invertebrate paleontology as well, particularly with arthropods (which includes trilobites), molluscs (which includes ammonites), and echinoderms (for example sea lilies). Through the 1810s a number of conchologists like Brocchi or Sowerby contributed to the systematic description of fossil molluscs. Lamarck gained a reputation for the classification of invertebrates. By the early 1820s further classic monographs appeared, such as Brongniart's work on trilobites, d'Orbigny's work on cephalopods, or Miller's study of crinoids. In a conscious attempt to emulate, in the field of paleobotany, what Cuvier had done for fossil vertebrates, Adolphe Brongniart wrote a treatise on fossil plants (never finished), complete with a *Prodrome* analogous to Cuvier's earlier *Discours préliminaire*. This rapidly accumulating knowledge of fossil invertebrates and plants appeared, by the end of the 1820s, to amount to a powerful tool for the delineation, subdivision, and correlation of formations. The invertebrate fossils proved to be an even more valuable criterion of relative age than those of vertebrates.

The main advocate in England of the use of fossils in stratigraphy was John Phillips, who followed in the footsteps of his uncle, William Smith. Phillips's *Illustrations of the Geology of Yorkshire* (1829) provided many examples of the precise and detailed study of fossils in the context of their geological age. By around 1830 the use of fossils to indicate relative age had become so well established that Sedgwick commented, in his anniversary address to the Geological Society: 'Each succeeding year places in a stronger point of view the importance of organic remains, when we attempt to trace the various periods and revolutions in the history of the globe.' Perhaps with Fitton's lament of the previous anniversary in mind, he

---

[22] *Fossil Remains of the Animal Kingdom*, 1830, 376. See also Higgins, *Book of Geology*, 1842, 271–2.

continued: 'I do not deny the importance of mineralogical characters; I only mean to assert that, taken by themselves, they are no
certain indication of the age of any deposit whatsoever.'[23] The
following year Sedgwick stressed anew the importance of fossils as
indicators both of relative age, and of environment of deposition
(facies, in modern parlance). From the presidential chair at the
Geological Society this view echoed through a plethora of general
science magazines right through to the annual *Arcana of Science and
Art*. Similar views were worked out in de la Beche's *Geological
Manual* (1831) and in several books by Phillips in which he made a
speciality of statistical tables that showed the distribution of
different fossil types and groups through the succession of
formations. In his *Guide to Geology* (1834) Phillips stated that 'each
system of strata may be identified through its whole course, and
discriminated from the older and more recent systems, by a
judicious examination of a sufficient number of its organic
contents.'[24]

The idea of dating or correlating strata by means of fossils had
been around for at least two decades; but only by about 1830 was
enough systematic paleontology known to make the idea applicable
in practice, long after a column of formational succession based on
lithology had been established.

The occurrence of distinctive assemblages of fossils, separated
above and below from different organic assemblages in the
succession of formations, strongly reinforced an interpretation of
earth history as a succession of worlds, each populated by a characteristic fauna and flora. Buckland's work contributed significantly
to the reconstruction of some of these worlds. Apart from his cave
work and the identification of the megalosaur, he added to the
paleontology of the iguanodon (1829).[25] More substantially,
Buckland described a particularly well-preserved specimen of yet
another monster from the Lias of Lyme Regis found by Mary
Anning. It was a flying reptile, the pterodactyl, specimens of which
were rare and only previously known from Solnhofen.[26]

The pterodactyl added sensationally to the inventory of geological monsters which were composed of seemingly disparate

[23] *Proc. Geol. Soc.* i (1834), 204; see ibid., 295.
[24] 69.
[25] 'On the Discovery of the Bones of the Iguanodon', *Proc. Geol. Soc.* i (1834), 159–60.
[26] Quenstedt presented a brief history of *Pterodactyli* discoveries in his *Über 'Pterodactylus suevicus'*, 1855.

anatomical elements. Buckland's imaginative description restored the pterodactyl to a creature which:

> somewhat resembled our modern bats and vampyres, but had its beak elongated like the bill of a woodcock, and armed with teeth like the snout of a crocodile; its vertebrae, ribs, pelvis, legs, and feet, resembled those of a lizard; its three anterior fingers terminated in long hooked claws like that on the fore-finger of the bat; and over its body was a covering, neither composed of feathers as in the bird, nor of hair as in the bat, but of scaly armour like that of an Iguana;—in short, a monster resembling nothing that has ever been seen or heard-of upon earth, excepting the dragons of romance and heraldry. Moreover, it was probably noctivagous and insectivorous, and in both these points resembled the bat; but differed from it, in having the most important bones in its body constructed after the manner of those of reptiles. With flocks of such-like creatures flying in the air, and shoals of no less monstrous Ichthyosauri and Plesiosauri swarming the primaeval lakes and rivers,—air, sea, and land must have been strangely tenanted in these early periods of our infant world.'[27]

Buckland surpassed his contemporaries in showing a unique talent in the reconstruction, not so much of the anatomy of various animals, but of their life habits and environment. Buckland was a paleoecologist long before Dollo. Paleoecology was at the heart of the restoration of former worlds, of landscapes of the geological past. Buckland's *Reliquiae Diluvianae* depicted the last world but one, when hyenas and cave bears preyed across the northern regions of Europe. An even earlier world had begun to emerge from the dawn of earth history, one in which gigantic reptiles roamed the earth, a period referred to by Mantell as 'the Age of the Reptiles'.[28] Buckland helped to reconstruct this long extinct world.

## SCENES OF ANCIENT WORLDS

Three geological phenomena in particular formed the basis for Buckland's paleoecology: fossil excrement or coprolites, animal footprints in sandstone, and fossil plants in a soil layer. Coprolites occurred in large numbers in Lyme Regis, where they were known as bezoar stones. They are dark grey and resemble in size and shape

[27] 'On the Discovery of a New Species of Pterodactyle', *Trans. Geol. Soc.* iii (1835), 217–18.

[28] 'The Geological Age of Reptiles', *Edinburgh New Phil. Journ.* Apr. to Oct. 1831, 181–5. See also id., *Wonders of Geology*, ii (1838), 379–80.

elongated pebbles or potatoes. Their origin, like that of so many concretions, seemed entirely enigmatic. Buckland, however, convinced his contemporaries that the bezoar stones were the fossil excrement of ichthyosaurs.

He based his coprolitic theory on a number of observations (figure 9). Chemical analyses showed that the coprolites consist of *album graecum* containing a large percentage of phosphate of lime, like fossil hyena excrement in caves; from this it followed that the basis of the coprolites had been bone. Polished sections revealed remnants of bones, fish scales, and pieces of cuttle-fish. In a number of nearly complete skeletons of ichthyosaurs coprolites occur within the ribs and near the pelvis suggesting that they had not yet been voided at the moment of death. He reasoned, most ingeniously, that the internal and external appearance of the coprolites indicated their passage through intestines like those of modern dog-fish or shark. Coprolites are arranged in a spiral whorl around a central axis; in addition their surface may show what looks like the impression of a vascular texture. The intestines of dog-fish and shark have their inside surface enlarged by a spiral valve which winds around the interior; Buckland injected the intestines of two common species of English dog-fish with cement and managed to reproduce the internal and external coprolitic features.

Buckland described coprolitic substances from as early as the Carboniferous age right up to modern guano. Coprolites extended the study of anatomy from fossil bones to the soft tissues of intestines, even though the latter had not been preserved. More interestingly, the composition of coprolites formed direct evidence of what extinct animals ate. Ichthyosaurs were not only carnivorous, but cannibalistic; some of the bones in their coprolites belonged to their own young. These coprolites showed that carnivorous habits had existed for a very long time, long before the existence of man and of human sin. Buckland wrote:

In all these various formations our Coprolites form records of warfare, waged by successive generations of inhabitants of our planet on one another: the imperishable phosphate of lime, derived from their digested skeletons, has become embalmed in the substance and foundations of the everlasting hills; and the general law of Nature which bids all to eat and be eaten in their turn, is shown to have been co-existensive with animal existence upon our globe; the *Carnivora* in each period of the world's history

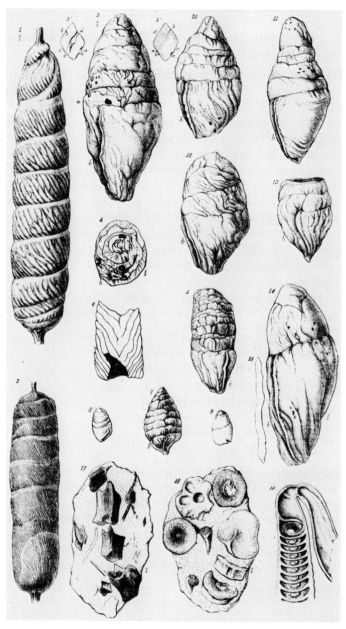

9. Artificial coprolites (1, 2), fossil coprolites from Lyme Regis (3 to 15), intestinal section of a shark (16), and sectioned coprolites with identifiable, undigested food particles (17, 18).

fulfilling their destined office,—to check excess in the progress of life and maintain the balance of creation.[29]

This work is typical of Buckland's predilections. It was a relatively small piece of work, but extremely ingenious; it had some economic value in that coprolitic accumulations turned out to be useful as fertilizer in agriculture; and fossil excrement appealed to Buckland's rather coarse sense of humour.

The study of coprolites turned into something of a craze. Suddenly an intriguing new perspective of previously misidentified fossil objects had opened up. Those interested in chemistry published analyses of the composition of coprolites; among them were Daubeny, Prout, and Wollaston.[30] Others made comparative observations of the faeces of fish and reptiles, and how they void their excrement. One correspondent admiringly called Buckland 'the greatest discoverer of this pre-eminent age of discovery'.[31]

Coprolites were an easy target for jest. De la Beche drew a humorous sketch which depicted Buckland's 'coprolitic vision'.[32] Duncan versified the following advice to an Oxford student:

> Approach, approach ingenuous youth
> And learn this fundamental truth
> The noble science of Geology
> Is bottomed firmly on Coprology
> For ever be Hyenas blessed
> Who left us the convincing test
> I claim a rich *coronam auri*
> For the thesauri of the Sauri.[33]

Duncan suggested also that the vice-chancellor choose the subject of coprolites for the Latin verse prize. The subject undoubtedly added to the fun of contemporary natural history. Frank Buckland delighted, in his *Curiosities of Natural History*, in relating the following: 'I have seen in actual use ear-rings made of polished portions of coprolites (for they are as hard as marble); and while admiring the beauty of the wearer, have made out distinctly the

---

[29] 'On the Discovery of Coprolites', *Trans. Geol. Soc.* iii (1835), 235.

[30] E.g. Prout to Buckland, 2 Oct. 1829, OUM, Bu P. See also Buckland to Anstice, May 1829, FM, Bu P.

[31] E.g. Anstice to Buckland, 19 Jan. 1830, OUM, Bu P.

[32] Reproduced in McCartney, *Henry de la Beche*, 1977, 49.

[33] Coprolite file, OUM, Bu P. A slightly different version occurs in Frank Buckland, *Curiosities of Natural History*, 2nd series, 1893 edn., 6.

scales and bones of the fish which once formed the dinner of a hideous lizard, but now hang pendulous from the ears of an unconscious *belle*, who had evidently never read or heard of such things as coprolites.'[34]

From the late 1820s Buckland's name became associated with the study of fossilized animal footprints and tracks. This subject of ichnology began in the summer of 1827 when a Henry Duncan wrote to Buckland for his opinion on impressions found on slabs of sandstone in a building-stone quarry of Corn Cockle Muir near Dumfries in the south of Scotland.[35] The slabs were of New Red Sandstone (Permo-Triassic), much older than the rocks in which the majority of reptilian fossils had been found. Buckland concluded that the impressions were genuine, and that they had been produced by a quadruped walking on the sand before subsequent processes had caused it to lithify.

One night, long after midnight, Buckland suddenly hit upon the idea that a tortoise might have produced the sandstone impressions. He called his wife down to cover the kitchen table with paste, while he fetched the pet tortoise from the garden and forced it to walk over the moistened flour. 'The delight of the scientific couple may be imagined when they found that the footmarks of the tortoise on the paste were identical with those on the sandstone slab.'[36] Some of Buckland's colleagues such as Murchison and Sedgwick remained sceptical, but during a memorable meeting of the Geological Society Buckland repeated his tortoise experiment. The *London Magazine* reported that 'the whole geological world has been in raptures' at Buckland's tortoise theory, and it gave the following satirical account of the repeat experiment:

Everything being ready for the demonstration, and the interest of the scientific company wound up to the highest pitch, the tortoise was placed on the chalk, and, first of all, he flatly refused to stir a step. The members, upon this, very properly waxed impatient, got in a rage, and began kicking and banging him about, and maledicting him in an extremely moving manner. They had much better, however, have refrained from these stimulants; for when the tortoise was at last prevailed on to walk, he insisted on walking as straight as an arrow; whereas the antediluvian tortoise's

[34] Ibid., 8.
[35] Duncan to Buckland, 11 June 1827, plus several later letters, OUM, Bu P. See Buckland to Featherstonhaugh, 8 Sept. 1827, CUL, Se P; and Buckland *Geology and Mineralogy*, i. 260–6.

march was as crooked as a ram's horn! Various arguments, however, were used to console them. It was suggested that the tortoise might have forgotten the true manner of walking while confined in the ark; and that owing to this circumstance the proper step might have been lost by his descendants. Or it might be, that chastened by the deluge, his slow race had returned to the path of rectitude, which they had, in the universal degeneracy, wilfully deserted for devious ways. Or perhaps, they had one way of walking on red sandstone, and another of soft chalk: one manner in private, and another before scientific beholders. Or, probably, the march of mind might be the cause; and tortoises, quicker than Tories, may have rejected the maxim, *Stare in antiquas vias*, and studied, like utilitarians, the shortest means to the proposed end.[37]

In 1833 the imprints of a large animal were discovered in a New Red Sandstone quarry of Hessberg near Hildburghausen in Thüringia. Because these impressions looked like those of human hands J.-J. Kaup coined the name *Chirotherium* for the corresponding animal.[38] In 1836 a further discovery of footprints was made in the New Red Sandstone of the Connecticut Valley by the American geologist Edward Hitchcock.[39] He interpreted the gigantic, three-toed imprints as those of large birds, though they belonged in fact to bipedal dinosaurs. Buckland included all these footprint discoveries in his Bridgewater Treatise. The significance of the New Red footprints was that, although no reptilian bones or teeth had yet been found, the existence of reptiles could still be inferred and such features as their length of stride be measured.

In 1838 a sensational find of both tortoise tracks and *Chirotherium* impressions was made in a New Red Sandstone quarry on Storeton Hill near Liverpool. John Cunningham, a local architect and member of the Liverpool Natural History Society, reported the discovery to Buckland, who in his turn brought it to the attention of the British Association.[40] The sandstone slabs also exhibited ripple

[36] Gordon, *Life and Correspondence of Buckland*, 217.

[37] Anon., *London Mag.* x (1828), 360–1.

[38] See Krämer and Kunz, '*Chirotherium,* das "unbekannte" Tier', *Natur und Museum* xcvi (1966), 12–19. *Cheirotherium* is the correct spelling, but *Chirotherium* has taxonomic priority.

[39] 'Description of the Foot Marks of Birds', *Am. Journ. Sci.* xxix (1836), 307–40.

[40] Cunningham to Buckland, 5 Sept. 1838, plus many later letters, OUM, Bu P. Buckland, 'Account of the Footsteps of the Cheirotherium', *Report BAAS, 1838,* Trans. Sect. 85. See also Sarjeant, 'History and Bibliography of the Study of Vertebrate Footprints in the British Isles', *Palaeogeogr. Palaeoclim. Palaeoecol.* xvi (1974), 265–378.

marks and impressions of raindrops, a feature which led Buckland to speculate about the first known 'fossil shower'.[41]

Coprolites and the fossil tracks of animals were evidence of periods of geological quiet, during which animals had lived, consumed food, and swam or walked around. Buckland's most convincing proof of such periods, apart from his hyena den study, was an ancient dirt bed interpreted as an *in situ* layer of plant growth or fossil soil. In 1828 Buckland read to the Geological Society a paper 'On the Cycadeoideae, a Family of Fossil Plants Found in the Oolite Quarries of the Isle of Portland'.[42] Some of the plants from the Oolite were coniferous, but others were identified by Buckland as cycadeous, closely allied to the living genera *Cycas* and *Zamia* commonly seen in greenhouses. He concluded that they were proofs of the existence of a tropical climate at the place and time of their original growth.

In his Weymouth paper (read in 1830) Buckland proceeded to show that the fossil Cycadeoideae had actually grown in the area where they were found, namely the Portland 'Dirt Bed' described by Webster and subsequently also by Fitton.[43] The bed is about one foot thick, composed of black, lignitic mould; in it silicified stems of conifers occur, some prostrate and some erect; trunks of Cycadeoideae are also found in an upright position, their roots lodged in the black mould and evidently in position of growth. By this interpretation the dirt bed is a fossil soil layer with some of the trees *in situ*:

We have a measure of the duration of this forest, in the thickness of decayed vegetable matter and soil, which has accumulated more than a foot of black earth around the roots of these trees. The regular and uniform preservation of this thin bed of black earth over a distance of so many miles, shows that the change to the next state of things was quiet and gradual; since the trees that lie prostrate on this black earth would have been swept away had there been any violent agitation, or sudden irruption of waters.[44]

The lucid presentation and impeccable logic of this work turned it into yet another classic of the period. The two illustrations of the

[41] *Abstracts Proc. Ashmolean Soc.* i (1844), no. xvi, 5–7. See also Buckland's anniversary address to the Geological Society of 21 Feb. 1840, *Proc. Geol. Soc.* iii (1842), 245–7.

[42] *Trans. Geol. Soc.* ii (1829), 395–401.

[43] See Fitton, 'Observations on Some of the Strata between the Chalk and the Oxford Oolite', *Trans. Geol. Soc.* iv (1836), 103–389.

[44] Buckland and de la Beche, 'On the Geology of the Neighbourhood of Weymouth', ibid., 16.

dirt bed in the Weymouth paper became widely used in geology text of the nineteenth century. The quality of this work was in part the result of Buckland's interest in contemporary analogues of the paleontological past. This kind of actualism characterized his work and permeated his Bridgewater Treatise. Examples abound: he interpreted a black substance found in the Lias of Lyme Regis as the fossil ink-bag of an ancient cuttle-fish, and he took delight in having the fossil specimens drawn in their own reactivated ink.[45]

These and a variety of other discoveries of fossil crocodiles, tortoises, etc., especially from the Jurassic limestones in England, inspired de la Beche to sketch an idealized view of *Duria Antiquior*, Dorset at the time when the Liassic monsters inhabited the region (figure 10). The picture centres around a number of ichthyosaurs and plesiosaurs, but it also depicts pterodactyls, cycadeoideae, ammonites, etc. The animals are involved in a variety of activities, fighting, hunting, eating, or defecating.

The picture, drawn around 1830, is the first of its kind. Buckland was thrilled by it, and complimented de la Beche profusely. Copies were made and circulated at home and abroad.[46] Buckland used to keep a good supply for his lectures, 'in order to bring to the minds of his audience the reality of the subjects on which he had been conversing'.[47] Much in such restorations of land- and seascapes was bound to be conjectural. Buckland did not include *Duria Antiquior* in his Bridgewater Treatise, even though he did venture a modest 'Imaginary Restoration of Pterodactyles' which showed some of the animals in flight, others hanging off a cliff face, and a cycad with a giant dragon-fly.[48]

John Martin, also, made illustrations of two scenes from the extinct past. One was a landscape, *The Country of the Iguanodon*, executed for Mantell's *The Wonders of Geology* (1838). The other was a seascape centered on an ichthyosaur fighting plesiosaurs for Hawkins's *Book of the Great Sea Dragons* (1840). The paleontological quality of Martin's work was distinctly inferior to that of de la Beche. The artistic idiom of both men can be recognized in a

---

[45] Buckland, 'On the Discovery of a Black Substance Resembling Sepia', *Proc. Geol. Soc.* i (1834), 97–8.

[46] Buckland to de la Beche, 'I have a capital class which I am sure is 30 per cent better for your Duria Antiquior by way of a syllabus.' 1 May 1831, NMW, Bu P.

[47] Frank Buckland, *Curiosities of Natural History,* 2nd series, 1893 edn., xi.

[48] *Geology and Mineralogy,* ii, pl. 22 P.

10. Henry de la Beche's *Duria Antiquior* or ancient Dorset with ichthyosaurs, plesiosaurs, pterodactyls and other extinct animals obeying the 'law of nature which bids all to eat and be eaten in their turn'.

number of similar land- and seascapes from 'the Age of the Reptiles' which appeared during the 1840s and even later.

New and original illustrations appeared as well. Franz Unger's *Die Urwelt* (1847) presented a series of high-quality drawings in which special attention was paid to the fossil flora. More derivative were wall charts like Perrot's *Tableau du monde antédiluvien*. Benjamin Waterhouse Hawkins made a number of life-size models of extinct monsters to be exhibited in the Crystal Palace grounds at Sydenham in Greater London,[49] and a dinner was held inside an Iguanodon model, attended by leading scientists. The models of the aquatic monsters were superb; a drawing of these by Waterhouse Hawkins was added to the posthumous edition of Buckland's Bridgewater Treatise. The dinosaur models, however, were still rather inaccurate. Despite Cuvier's belief that a single bone might suffice to reconstruct an entire animal, an exact representation of the dinosaurs was only possible after the discovery of more complete skeletons, such as the iguanodons of Bernissart in southern Belgium.[50]

[49] See Owen, *Geology and Inhabitants of the Ancient World,* 1854. A drawing of the monsters by Waterhouse Hawkins was added to Buckland's *Geology and Mineralogy*, 3rd edn. of 1858, ii, pl. 23.

[50] See Casier, *Les Iguanodons de Bernissart,* 1960.

# 12
# Progressive Succession among Fossils

## CENTRAL HEAT

Thus far two important discoveries of historical geology have been discussed: first, that rock formations occur in an orderly sequence and form a relative time-scale from old to young; second, that individual formations or groups of these may be characterized by assemblages of fossils and are the record of periods of earth history. A skeleton column of international rock succession existed by around 1820. The reconstruction of some former worlds was accomplished during the 1820s, most sensationally so for 'the age of the reptiles'.

A third major aspect of historical geology concerned the relationship of one fossil assemblage to the next. How should one view the sequence of former worlds? Was there a linked sequence or a law of organic succession? More specifically, what characterized the relationship between the earlier world during which the reptiles reigned and the later age of mammalian domination?

In the course of the first few decades of the nineteenth century a consensus emerged that the relationship of successive fossil worlds was one of progress or progressive development. Progress through earth history was defined by reference to two criteria, one taxonomic, the other ecological. Taxonomic progress meant that the lower and simpler forms of life appear earlier in the geological succession and the higher and more complex ones later, as for example in the sequence extending from reptiles in the Secondary rocks to mammals in the Tertiary.

The ecological criterion linked organic succession to environmental change, and progress was defined as the successively increased habitability of the earth to higher forms of life and in particular to man. This criterion reduced taxonomic progress to a subsidiary effect of environmental change. In its most sophisticated form the environment was defined not merely by physical factors such as temperature, humidity, or atmospheric composition, but also by animal and plant communities. Thus the unit of progressive development was, in modern parlance, the ecosystem.

The ecological criterion connected paleontology to the study of the earth as a planet. During the early part of the nineteenth century there was a revival of the old notion of a central heat, which stated that the earth had originated as an incandescent mass, that this mass had cooled down gradually and acquired a solid crust, and still retained a core of primeval heat. The central heat was believed to have influenced the climate of the earth, especially during its early stages of thermal evolution. The ecological theory of progressive development was given a ready-made context of change in the concept of a central heat and its corollary of climatic amelioration. The dominant form of life during a particular period of earth history had been the one most perfectly adapted to contemporary environmental conditions. Cold-blooded reptiles reigned during the Secondary when the climate was believed to have been stable and warm; the domination by mammals came during the Tertiary, because by that time the earth had cooled down sufficiently for there to be seasonal variations, to which warm-blooded mammals were perfectly adapted.

The theory of progressive development became popular throughout Europe. Early advocacy of it was restricted to continental Europe, but during the 1820s and 1830s the idea of progressive succession became an integral part of the geology of the English school. There were substantial differences in the ways in which the theory was developed and culturally integrated in England, Germany, and France. In Germany it interacted with *Naturphilosophie* and transcendental idealism which favoured the taxonomic definition of fossil progress; in France much credence was given to the theory of a central heat because of its association with the exact sciences. The English school was barely familiar with German idealism and suspicious of the primarily deductive arguments for a central heat, but instead, by virtue of its interest in natural theology, embraced the ecological theory of progressive development with its emphasis on adaptation.

Speculation about the nature of the fossil succession began as soon as the Wernerians had worked out the beginning of a stratigraphic column. Werner himself was interested in fossils and their geological occurrence.[1] As early as 1806 von Buch made progressive development ('das Fortschreiten der Bildungen in der Natur') the subject of his inaugural lecture to the Königliche Akademie der Wissenschaften of Berlin.[2] Progressive succession began to acquire a solid evidential basis with Cuvier's study of fossil quadrupeds. He observed that the higher classes of quadrupeds appear later in the geological record, and that aquatic forms appear before terrestrial ones.[3] The late appearance of mammals and the apparent absence of human fossils were, in particular, interpreted as evidence of progress.

At this stage hasty generalizations were easily made. Parkinson interpreted the progressive succession of fossils as analogous to the order of creation in the first chapter of Genesis, i.e. from plants to water animals to land animals and man. This analogy underscored the anthropocentric nature of the theory of geological progress.[4]

Idealized schemes of both formational and fossil succession met with the disapproval of such staunch empiricists as Greenough. In his *First Principles of Geology* (1819) he argued against a progressive succession of fossils, citing a variety of anomalous complexities in the paleontological record.[5] Admittedly not enough was known at this time to regard a progressive nature of earth history as definitively established. This was recognized on both sides of the Channel.[6]

In England a cautious opinion was expressed by young Lyell in 1826, who at this time had not yet assumed his full Huttonian identity. In a review for the *Quarterly Review* of the *Transactions of the Geological Society*, Lyell summarized those facts which by the mid-1820s were regarded as well established, e.g. that the formations composing the earth's crust are not thrown together in inexplicable confusion, but arranged in a regular order of superposition which is never inverted; that groups of strata are often characterized by particular assemblages of organic remains; that in the oldest rocks no impressions of plants or animals have been discovered; that the fossils from the oldest strata differ most widely from contemporary organisms; that in ascending order from the

[1] According to Jameson, in his notes to Cuvier's *Theory of the Earth*, 1813, 225–7.
[2] *Leopold von Buch's Gesammelte Schriften*, ed. Ewald, Roth, and Eck, ii (1870), 4–12.
[3] *Theory of the Earth*, 105–14.
[4] See Parkinson, *Organic Remains of a Former World*, iii (1811), 455.
[5] 281–4.
[6] E.g. Humboldt, *Geognostical Essay*, 1823, 45–6.

early to the more recent deposits fossil forms look increasingly like species now alive; and that fossil plants and animals belonging to families and genera now confined to the tropics abound in the strata of high latitudes.[7]

The last fact suggested that a change of climate, as had been inferred from the theory of a central heat, had indeed occurred. The theory had its antecedents in the cosmogonies of Descartes and Leibniz; Buffon had advocated it in his widely read *Époques de la nature* (1778). Even though such cosmogonical speculations had been abjured by most geologists of the early nineteenth century, the notion of a central heat gained new respectability as a result of a combination of inferences from astronomy, mathematics, and physics, particularly in France.

The idea that the earth had begun in a state of incandescence appeared to be supported by the shape of the earth, an oblate spheroid flattened at the poles. Laplace's nebular hypothesis enhanced its credibility. Fourier's work on the theory of heat conferred on it the prestige of mathematics and physics.[8] But Cordier connected the theory to geological observations. In his classic 'Essai sur la température de l'intérieur de la terre' (1827) he listed a large number of measurements of the thermal gradient of the earth's crust.[9] Various phenomena were brought to bear on the possible existence of a central heat. First, there was the increase of temperature in going down a mine shaft. Temperature readings were taken of air, water, and occasionally the rock wall of mines, in order to substantiate the thermal gradient. Second, there was the similar gradient in wells and springs. The famous artesian well at Grenelle in Paris was sunk during the 1830s, a project to which both Arago and Cordier contributed and which provided an opportunity to document the thermal gradient below Paris. Third, volcanism was explained as a form of igneous activity generated by the hot core of the earth.

Adolphe Brongniart presented a particularly convincing case for climatic change in his *Prodrome* (1828) on fossil plants. He grouped these according to their geological age and drew up a number of statistical tables which showed that plants, like animals, exhibited

[7] xxxiv (1826), 507–8.
[8] See Lawrence, 'Heaven and Earth', *Cosmology, History, and Theology*, ed. Yourgrau and Breck, 1977, 253–81.
[9] *Mém. Acad. Sci.* vii (1827), 473–556.

progressive development through earth history.  Early in the coal period, primitive but gigantic ferns and other cryptogamous plants dominated; later, there was a preponderance of gymnosperms such as conifers and cycads; later still, the angiosperms or flowering plants took over as the ruling type of vegetation on earth: 'Nous pouvons donc admettre parmi les végétaux, comme parmi les animaux, que les êtres les plus simples ont précédé les plus compliqués, et que la nature a créé successivement des êtres de plus en plus parfaits.'[10] Brongniart argued that the giant cryptogams of the coal period flourished because of tropical heat and humidity.  Such conditions, he believed, had been produced by internal heat, and the later changes in plant life marked not only a climatic cooling, but also a decrease in the carbon content of the atmosphere and an increase in the expanse of dry land:

Il est bien peu de physiciens qui doutent maintenant que la terre n'ait eu, dans les premiers temps de sa formation, une température plus élevée que celle dont elle jouit actuellement.  La nature et la grandeur des végétaux du terrain houiller présentent une des confirmations les plus fortes de cette théorie que la géologie puisse fournir; et la diminution successive de cette température est sans aucun doute une des causes qui ont le plus influé sur les changements que la végétation a subis depuis cette époque reculée jusqu'à nos jours.[11]

By the end of the 1820s the theory of a central heat had become in France the backbone of a widely held belief in progressive earth history.  Central heat was suggestively depicted by Nérée Boubée (1833) on his popular wall chart which showed 'l'état du globe à ses différens âges', or the thermal evolution of the atmosphere and crust of the earth (figure 11) by process of secular cooling.[12]

The English school treated the theory of a central heat with circumspection.  Others rejected the concept altogether.  Its primarily deductive basis did not agree with the Baconian

---

[10] 'We can thus accept that among plants, as among animals, the simplest beings have preceded the most complex, and that nature has successively created increasingly perfect beings.' 221.

[11] 'There are very few physicists who are not now convinced that the earth has had a higher temperature during the first period of its formation than that which it enjoys at present. The type and size of the Carboniferous plants constitute one of the strongest confirmations of this theory which geology can offer; and the gradual diminution of that temperature is without any doubt one of the most influential causes of the changes which the vegetation has undergone from that remote period till today.' 222.

[12] French and German wall chart editions existed; the third French edition was published by Boubée as the frontispiece to his *Géologie élémentaire*, 1833.

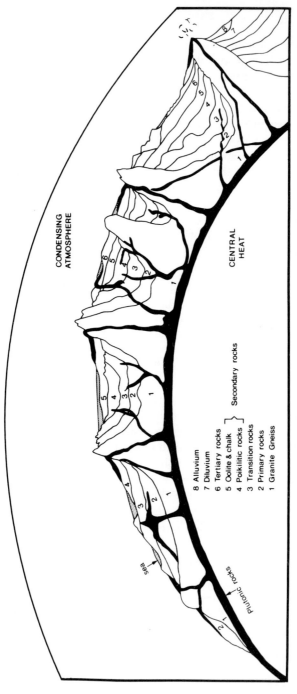

CONDENSING
ATMOSPHERE

CENTRAL
HEAT

sea

plutonic rocks

8 Alluvium
7 Diluvium
6 Tertiary rocks
5 Oolite & chalk
4 Poikilitic rocks
3 Transition rocks
2 Primary rocks
1 Granite Gneiss

Secondary rocks

11. The evolution of the earth's atmosphere and crust by secular cooling, according to Nérée Boubée and redrafted from his coloured wall chart.

empiricism of the Geological Society, and the consanguinity of the
theory with the despised cosmogonies of the eighteenth century was
not entirely counterbalanced by its new prestige in France. The
most scathing attack on central heat was launched, predictably, by
Greenough in an anniversary address to the Geological Society.[13]
Lyell rejected the theory because of its corollary of linear, progressive change.[14] Lindley and Hutton rejected Brongniart's evidence
for climatic change and taxonomic progress.[15] England's own
foremost astronomer, Herschel, suggested to the Geological Society
that climatic change may have been produced by a secular decrease
in the eccentricity of the earth's orbit rather than by heat loss.[16]

The main reason for the wariness of the English school regarding
a central heat was probably Daubeny's commitment to the
chemical theory of volcanism. Daubeny's stature as a scientist was
modest, but his influence was pervasive. Early in the century Davy
had suggested that chemical reactions in the earth's crust may
produce the heat which causes volcanic eruptions. This chemical
view did away with the connection between central heat and
volcanism. Davy later abandoned the theory, but Daubeny held on
to it. The relationship between chemistry and the study of
volcanism was so intimate that at Oxford the lectures on volcanoes
were not given by Buckland, but came under the jurisdiction of
Daubeny as professor of chemistry. In the published version of
these lectures he reaffirmed his belief in the chemical theory of
volcanic heat production. Daubeny attributed the heat in deep
mines to local causes, e.g. the decomposition of pyrites, and the
warmth of miners, their horses, or their lights.[17] During the 1830s
he took issue with Karl Gustav Bischoff, who presented a powerful
argument in favour of a central heat.[18]

---

[13] *Proc. Geol. Soc.* ii (1838), 61–7.

[14] *Principles,* i (1830), 141–3.

[15] *Fossil Flora of Great Britain,* i (1831), xiv-xx.

[16] 'On the Astronomical Causes which May Influence Geological Phenomena', *Trans. Geol. Soc.* iii (1835), 293–9.

[17] *Description of Active and Extinct Volcanos,* 1826. See Anon., 'Dr. Daubery on Volcanos', *Ann. Phil.* xii (1826), 215–26; see also ibid., xi (1826), 259–60; v (1823), 33–43.

[18] Bischoff's work was serialized in the *Edinburgh New Phil. Journ.* e.g. 'On the Natural History of Volcanos and Earthquakes', xxvi (1839), 25–81, 347–86. Daubeny criticized, ibid., 291–9, xxvii (1839), 158–60. Bischoff responded, ibid., xxx (1841), 14–26. Bischoff's book, begun as an answer to a prize question by the Hollandsche Maatschappij in 1832, was translated into English: *Physical, Chemical and Geological Researches on the Internal Heat of the Globe,* 1841.

In spite of Greenough's empiricism, Lyell's Huttonian stance, and Daubeny's chemical bias, the English school followed the continental lead on central heat. De la Beche in his *Geological Manual* (1831), Conybeare in his *Report* (1832), Sedgwick in his *Discourse* (1832), and Phillips in his *Guide to Geology* (1834) successively came out in cautious support for the theory. Characteristic is de la Beche's prefatory comment to his *Researches in Theoretical Geology* (1834): 'Although the theory of central heat and the former igneous fluidity of our planet have been much dwelt upon in the following pages, the author trusts that he will not be considered so attached to these views as not to be ready to reject them and embrace others which may afford a better explanation of an equal number of observed facts, should such be brought forward.'[19]

Buckland also believed in a central heat and recognized its explanatory power as a theory. But in his Bridgewater Treatise he presented the chemical theory as well as the notion of internal heat, not wishing to alienate his colleague Daubeny. Scrope sharply criticized this half-hearted position.[20] But Buckland's real support for the theory of a central heat is unquestionable. He used Boubée's wall chart for his lectures, and even depicted a minia-ture of it on proofs of the fold-out diagram of the earth's crust in his Bridgewater Treatise. Buckland intended to add a separate chapter to his Treatise 'On the internal Temperature of the Earth'. In the end he omitted this chapter, but fragments that exist in manuscript make his position clear:

The question of internal heat forms an important part of the physical history of the globe, and is so mixed up with the most probable theory that has been proposed to explain the figure of the earth and the origin of unstratified rocks, and so ultimately connected with the changes that have affected animal and vegetable life during the deposition of stratified rocks, that although the subject is as yet involved in much uncertainty, its mention cannot with propriety be omitted on the present occasion.[21]

Buckland was convinced of the truth of a central heat most of all because of its consequence of climatic change, which explained otherwise anomalous fossil occurrences like the presence of corals and plants indicative of a warm climate in northern latitudes and even in the polar regions: 'We can discover no method of referring

[19] v.
[20] *Quarterly Review*, lvi (1836), 37–8.
[21] Internal temperature file, OUM, Bu P.

these phenomena to the influence of the sun; but if we suppose the climate of the ancient world to have derived its greater and more uniform warmth from internal heat, we find an explanation of the gradual diminution of this temperature in the gradual radiation of heat from the earth's surface into space.'[22]

The acceptance of a central heat by the English school came relatively late. It was preceded by an indigenous, provincial tradition of mining geology which had appreciated the notion of internal heat as early as continental workers. The work by Robert Fox in Cornish mines had been especially significant. As early as 1815 Fox had instigated temperature measurements in mines, and a few years later had communicated the results to the Royal Geological Society of Cornwall.[23] During the early 1830s this provincial tradition joined forces with the English school when Fox was commissioned to report on the temperature of mines to the British Association.[24] Bakewell unreservedly promoted the idea of a central heat in the 1830s editions of his textbook.[25]

ORGANIC PROGRESS

Even though a central heat was only hesitantly accepted by the English school, its companion theory of progress, ecologically defined, was wholeheartedly embraced. An early expression of the English position appeared in the *Annals of Philosophy* (1825), written by Alexander Crichton, a physician of international repute:

When the character of the vegetables and animals of the ancient world is duly considered in a physiological point of view as testimonies of temperature, we are led to the belief that the various living forms appeared in regular succession accordingly as the temperature of the earth suffered diminution; each succeeding race becoming fitted by its peculiarity of organisation to support a colder climate, and increasing vicissitudes of heat and cold.[26]

The fullest exposition of the theory that organic progress is a function of environmental change was given by Davy in his curious

---

[22] Ibid.

[23] See Fox to Buckland, 11 Mar. 1832, OUM, Bu P

[24] 'Report on Some Observations on Subterranean Temperature', *Report BAAS, 1840*, 309–19.

[25] *Introduction to Geology*, 1833 edn., 561–2; 1838 edn., 6–7, 600–7.

[26] ix (1825), 211.

*Consolations in Travel* (1830). This was a fictional travelogue which made possible a free and unguarded expression of a scientific belief which, though widely held, had not yet been substantiated to the extent that it could be formally addressed to the Geological or Royal societies. Davy sketched an integrated perspective of earth history along the lines of secular cooling and organic progress. It included igneous activity as the driving force of change, chemical precipitation as a subsidiary cause of rock deposition, environmental change as the main condition of fossil progress, a decrease of the intensity of geological processes with time, and episodes of paroxysmal uplift, the last of which had emplaced the diluvial gravel shortly before the creation of man.[27] Davy's perspective, if not vision, of the past represented the majority opinion of the English school. It was frequently cited, both in England and abroad.

The paleontological refinement of this perspective of change and progress was primarily the work of Buckland. His work on cave hyenas, on coprolites, on cycads *in situ*, and on a number of other subjects, contributed to the study of fossils not just in the abstract context of their taxonomic position, but in the circumstances of their ancient habitat. To Buckland a fossil was more than a taxonomic entity; it represented a member of an ancient community from a specific period of earth history. He rejected the notion that progress is merely a climb up the taxonomic ladder, from a less perfect to a more perfect level of organization, and emphasized that perfection in this context is a relative notion which has little meaning in the abstract, but is a function of adaptation to a particular environment: 'Our ideas of perfection are all relative; absolutely each animal is equally perfect.'[28]

Buckland illustrated this belief, giving examples of perfect adaptation such as the very ancient trilobites, described in the delicate detail of their eyes, or the more recent megatheria. He was especially intrigued by the skeleton of the megatherium, because its ponderous and awkward appearance made it seem poorly designed; but to the delight of his contemporaries Buckland demonstrated that even this fossil could be seen to be perfectly constructed when viewed in the context of its ancient habitat of the South American pampas.[29]

[27] 133–7. During the 1830s alone the booklet went through five editions. See *Edinburgh New Phil. Journ.* Jan. to Apr. 1830, 320–2, where Davy's progressivism was reproduced.
[28] Miscellaneous paleontology file, OUM, Bu P.
[29] See ch. 18.

This position was inspired by a belief in divine design in nature; Buckland's Bridgewater Treatise which considered geology in its relationship to natural theology was in effect a grand argument for adaptation and design throughout earth history. Because of this emphasis on adaptation, progress appeared best defined by the habitability of the earth to successive organic communities. Buckland wrote 'that the course of nature has not been a series of experiments each successively improving on that which preceded it, but that from the beginning every organized being was created perfect with relation to the functions which it was destined to perform in the then existing state of the world. The fitness of the world for animal life appears to have been progressive.'[30] The epithet 'imperfect' would thus apply only to an earlier state of the earth hypothetically considered as a habitat for later forms of life.

Phillips explained the ecological theory of geological progress in his lucid and semi-popular style of writing; Owen developed it. He declared at the end of his 'Report on British Fossil Reptiles' (1841): 'The general progressive approximation of the animal kingdom to its present condition has been, doubtless, accompanied by a corresponding progress of the inorganic world.'[31] Owen emphasized the composition of the atmosphere as a determinant of organic progress. The reign of the reptiles had occurred because the reptilian respiratory system was better suited than that of mammals to the atmosphere at that stage of its evolution, containing as it did less oxygen and more carbon dioxide than it does now. Owen's 'Report' had considerable impact; Buckland marked the completion of its delivery to the British Association by eulogizing Owen as a worthy successor to Cuvier; and the *Literary Gazette* devoted more than twenty columns to its review.[32]

Cuvier's *Ossemens fossiles* had paid scant attention to fossil fish. The potential interest of this class of vertebrate fossils to geology and to the theory of progressive succession was considerable; as a separate class fish cover a very long stretch of geological history and occur long before the reptiles. Towards the end of his life Cuvier encouraged young Agassiz in the study of fossil fish; in England Agassiz was given financial support for his research. The British Association, which promoted through its research programme the

[30] Miscellaneous paleontology file, OUM, Bu P.
[31] *Report BAAS, 1841,* 202.
[32] 1841, 513–19.

twin theories of central heat and progressive development, awarded Agassiz a number of grants during the 1830s under the patronage of Buckland, Murchison, and Sedgwick.[33] Broderip expressed appreciation and encouragement for the golden boy of continental paleontology in a review for the *Quarterly Review* (1836).[34]

Agassiz introduced a successful, fourfold classification of fish, based on their scales, and demonstrated that their geological distribution was progressive. Two orders (ganoids and placoids) originated in the Transition rocks, and the other two (cycloids and ctenoids) appeared in the Cretaceous. Progress could thus be seen not only as between classes of vertebrates, but also as between orders within a class.[35] Agassiz's *Recherches sur les poissons fossiles* (1833–43) became a new focus of interest for paleontology. It inspired Buckland to publish on fossil fish,[36] and he included Agassiz's early results in his Bridgewater Treatise. For this, he received substantial help from his young protégé,[37] and the latter further obliged Buckland by translating his Bridgewater Treatise into German (1839).

In due course, however, it became apparent that Agassiz was something of a cuckoo in the English nest. His continental education made him prefer a strictly taxonomic definition of progress. This lifted the phenomenon of paleontological succession out of its concrete context of environmental change onto the abstract plane of taxonomic and even embryonic analogy. Agassiz regarded the succession of fish, reptiles, mammals, and man as an inherently coherent sequence which represented a premeditated, divine plan of creation. Each progressive stage was a modification of the previous type of organization, according to a logic of anatomical perfection, and predicted the next. The successive stages were as inherently related to one another as the developmental stages of embryonic growth. Agassiz believed that the geological history of a class is recapitulated in the embryonic stages through which a present-day representative passes. He thus defined

---

[33] See Buckland to Phillips, 9 Aug. 1839, DRO, Bu P. See also Morrell and Thackray, *Gentlemen of Science,* 1981, *passim.*

[34] lv (1836), 433–45.

[35] 'On a New Classification of Fishes', read 1834, *Proc. Geol. Soc.* ii (1838), 99–102.

[36] 'Notice on the Fossil Beaks of Four Extinct Species of Fishes', read 1835, *Proc. Geol. Soc.* ii (1838), 205–6; see also ibid., 687–8.

[37] Buckland to Agassiz (returned with answers), 3 Dec. 1834, RS, Bu C, no. 61.

progress by a triple parallelism of paleontological succession, order of taxonomic classification, and embryonic development.[38] The contrast of Agassiz with Buckland is illustrated by the many footnotes which the former added to his translation into German of the latter's Bridgewater Treatise. Agassiz declared himself 'um die theologisch-teleologische Auslegung mancher Thatsachen etwas verlegen'.[39] To Agassiz perfection in fossils was not a function of adaptation, but was determined by the degree of an animal's approximation to man. 'Diese Begriffe von Vollkommenheit und Unvollkommenheit sind sehr unbestimmt. Auf dieser Erde scheint es mir nur einen Masstab zu geben, um in diesem Sinne das Verhältniss der Organismen zu einander zu bemessen, nämlich die Annäherung zum Menschen, als dem Herrn der Schöpfung.'[40] Divine design in the geological past was to be seen not so much in environmental provisions, as Buckland suggested, but in a taxonomic logic which was increasingly perfected in the direction of its ultimate goal, man:

Die sprechendsten Beweise von einer Planmässigkeit in der Aufeinanderfolge der Veränderungen, welche die Erde betroffen haben, finden wir vielmehr in der Entwickelung des organischen Lebens, in der Beschaffenheit der zuerst auftretenden Thiere und Pflanzen und in der Art, wie die späteren sich an die früheren anschliessen, bis zum letzten Ziel der Schöpfung, dem Erscheinen des Menschen, der durch die eigenthümliche Umgestaltung der Wirbelthiere in immer gesteigerte Menschenähnlichkeit durch alle Formationen augenscheinlich verheissen wird.[41]

[38] E.g., 'On the Succession and Development of Organised Beings at the Surface of the Terrestrial Globe', *Edinburgh New Phil. Journ.* xxxiii (1842), 388–99.

[39] 'somewhat embarrassed by the theological-teleological interpretation of many facts'. *Geologie und Mineralogie* i (1839), vi. See also 124.

[40] 'These concepts of perfection and imperfection are very vague. It seems to me that on this earth only one criterion exists to determine the relationship of the organisms to one another according to these concepts, namely the degree of resemblance to man as the lord of creation.' 126.

[41] 'We find instead in the development of organic life the clearest proofs of design amidst the successive changes which the earth has undergone, in the constitution of the first occurring animals and plants and in the way the later organisms are connected to the earlier ones, up to the final goal of creation, man, whose appearance is evidently foretold in the course of all formations through the peculiar transformation of the vertebrates into constantly increased similarity to man.' 51. See also Lurie, *Louis Agassiz,* 1960, chs. 1 and 2; Gould, *Ontogeny and Phylogeny,* 1977, 63–8; Bowler, *Fossils and Progress,* 1976, chs. 2 and 3.

These abstract concepts of German idealism were entirely alien to Buckland's Anglican empiricism, but the Scot Hugh Miller, a Calvinist like Agassiz, borrowed the notion of recapitulation in fish. He became familiar with it during the meeting of the British Association in Glasgow (1840) where he met Agassiz, and he included it in his classic book on the *Old Red Sandstone* (1841).

Once an outline of progressive development had been sketched, new discoveries took on added significance, depending on whether they confirmed the theory, like Agassiz's fossil fish did, or contradicted the expected trend of organic progress. Anomalous fossils of plants, animals, and even man were reported. Witham, who pioneered the use of thin sections to study fossil plants, concluded that conifers occurred as early as the coal period.[42]

A protracted controversy developed over the discovery, in the Stonesfield slate quarries near Oxford, of the jaws of didelphys, a small marsupial mammal. The slate was believed to be of Jurassic age, i.e. the early part of 'the age of the reptiles', long before the Tertiary when mammals were believed to have originated. The first known specimens were in the collection of Broderip while he was a student at Oxford. One of the jaw fragments was bought from Broderip by Buckland who later described it in his paper on the megalosaur, and identified and named it on Cuvier's authority as didelphys, or the Stonesfield Opossum as it became popularly known. The apparently anomalous position of didelphys excited much interest. Bakewell suggested that the fossil jaws had been reworked, introduced into the slate at a later stage through fissures.[43] Others questioned the Jurassic age of the Stonesfield slate or the mammalian affinity of the fossil jaws. In 1825 Constant Prévost, whose theory of continuous submergence during much of geological history did not agree well with evidence for early terrestrial habitation, questioned in particular the assigned age of the Stonesfield slate.[44] The precise correlation with other Jurassic strata in England was indeed unsure, but, as Lyell pointed out in his review of 1826, its Jurassic age was not in doubt. A twin paper by Broderip and Fitton in 1828

---

[42] In his *Observations on Fossil Vegetables*, 1831, Witham supported Brongniart's progressivism; but in his *Internal Structure of Fossil Vegetables*, 1833, he turned around and sided with William Hutton.

[43] 'Facts and Observations Relating to the Theory of the Progressive Development of Organic Life', *Phil. Mag. Ann. Chem.* ix (1831), 33–7.

[44] 'Observations sur les schistes oolithique de Stonesfield', *Ann. sci. nat.* iv (1825), 389–417.

reaffirmed the mammalian character of the jaws and the Jurassic age of the Stonesfield slate.[45]

The controversy did not die. It flared up again a decade later in 1838 when Henri de Blainville suggested that the fossil jaws belonged to a cold-blooded animal, probably a reptile.[46] Others joined the fracas; Agassiz believed that the nature of the animal was ambiguous and proposed the name *Amphigonus*;[47] but Achille Valenciennes defended its mammalian character.[48] The controversy was prominently reported by Charlesworth in the *Magazine of Natural History*. In 1838 Buckland made a pilgrimage to Paris, with two specimens, to convince the doubters; upon his return he made the material available to Owen who in separate publications in 1838 and 1842 settled the matter in favour of a mammalian identification.[49] Broderip had called the animal *Didelphys Bucklandi*; Buckland injected an element of jest into the controversy when he renamed the beast *Botheratiotherium Bucklandi*.[50]

Buckland could be casual about the didelphys anomaly. To those, like Agassiz, who defined progress on a strictly taxonomic basis, the anomaly was very disturbing and could not be explained: 'Das Vorkommen dieser Knochen in jurassischen Schichten, ist eben so befremdent, und lässt noch kein Anschliessen dieser Formen an andere schon bekannte derselben geologischen Periode, oder auch nur der nächstfolgenden, zu.'[51] To Buckland, however, for whom progress existed in a succession of ecosystems, each characterized by the optimal adaptation of a particular vertebrate class, the existence of a small, primitive mammal during the reign of reptiles was as little of a problem as the present-day existence of

[45] Broderip, 'Observations on the Jaw of a Fossil Mammiferous Animal', and Fitton, 'On the Strata from whence the Fossil Described in the Preceding Notice was Obtained', *Zool. Journ.* iii (1828), 408–18. Prévost's objections had been communicated by the *Edinburgh Phil. Journ.*, and had elicited an early response by Mantell, ibid., Apr. to Oct. 1826, 262–5.

[46] 'Doubts respecting the Class, Family, and Genus to which the Fossil Bones found at Stonesfield should be referred', *Mag. Nat. Hist.* ii (1838), 639–54; see also ibid., iii (1839), 49–57.

[47] Buckland, *Geologie und Mineralogie*, ii. 2–3.

[48] 'Observations upon the Fossil Jaws from the Oolitic Beds at Stonesfield', *Mag. Nat. Hist.* iii (1839), 1–10.

[49] 'Report on the British Fossil Mammalia', *Report BAAS, 1842*, 58–62. See also Owen to Buckland, 12 Dec. 1838, OUM, Bu P.

[50] Buckland to Brougham, 14 Dec. 1838, UC, Br P.

[51] 'The occurrence of these bones in Jurassic beds is equally strange, and does not allow for these forms to be connected to others already known from the same geological period, or even to those which follow.' See n. 47.

small reptiles during the reign of mammals and man. Phillips explicitly stated that didelphys should not be a problem to 'the geologist who, in the full spirit of Cuvier, regards the systems of life as definitely related now, and at all past periods, to the contemporaneous physical conditions of the globe, and uses the remains of plants and animals as monuments and guides to a right knowledge of these conditions'.[52]

More disturbing to Buckland was the discovery of fossil monkeys in Tertiary beds of the Siwalik Hills. It seemed to crack the foundations of the unique position of man as the non-fossilized apex of progressive change. But Buckland kept his good humour when he wrote to Brougham: 'There is no escape from the facts. But monkeys are not men, though the reverse is not always true.'[53]

A human fossil anomaly existed in footprints found in ancient rocks near St. Louis along the banks of the Mississippi. An American explorer, Henry Schoolcraft, reported their existence in 1822 to Silliman's *Journal*.[54] Mantell drew attention to them in his *Wonders of Geology* (1838), and commented that their discovery 'has not, as yet, excited among scientific observers the attention which its importance demands'.[55]

A physician, David Owen, re-examined the footprints. In a lengthy article for the *American Journal of Science* 'Regarding Human Foot-Prints in Solid Limestone' (1842), he confirmed that the rocks were indeed ancient—he thought they were Silurian. Owen, however, suggested that the footprints were not genuine, but had been carved by Indians. His reasons were that the prints do not form a track, that they are a single instance, and that they look too detailed to be fossil. But Owen added: 'Lastly, and chiefly, because of the age, nature, and position of the rock, and because no human remains whatever have hitherto been discovered in any similar formation.'[56] He emphasized that the climate at this early stage in the history of the earth would have made it uninhabitable to man. This chief reason was a *petitio principii*, a form of circular reasoning which has given fundamentalists cause to continue using this and subsequent discoveries of human footprints in ancient rocks as an argument against progressive succession.[57]

---

[52] *Treatise on Geology*, i (1837), 97–8.
[53] Buckland to Brougham, 26 Nov. 1838, UC, Br P.
[54] 'Remarks on the Prints of Human Feet', *Am. Journ. Sci.* v (1822), 223–31.
[55] i. 65.
[56] xliii (1842), 28.
[57] E.g. Rusch, 'Human Footprints in Rocks', *Creation Research Soc. Quart.*, Mar. 1971, 201–13.

# 13
# The Age of the Earth

The theory of a central heat provided a basis for estimating the age of the earth. The stretch of time from the moment the earth had acquired a solid crust (Leibniz's 'consistentior status') till the present day could be calculated, given a figure for the rate of heat loss. Buffon had dated the earth at some 76,000 years of age.[1] Boubée gave dates on his chart which added up to over 300,000 years for the total duration of earth history.[2] This method of dating was perfected by Kelvin in the 1850s and subsequent decades.[3]

Herschel's alternative hypothesis to that of internal heat, namely a decrease in the eccentricity of the earth's orbit, also provided a basis for calculating the age of the earth. Herschel estimated that the decrease from an initially elliptical state to the current nearly circular orbit would have taken at least 600,000 years.[4]

These and other methods of geochronology were at the centre of scientific interest during the later nineteenth century. In the first half of the century, however, interest was focused not on age expressed in numbers of years, but on the qualitative evidence for a long stretch of earth history prior to the age of man. The age of the earth and the duration of the periods of its history could not be reliably estimated until the full sequence of rock formations and of fossil communities, the very events to be dated, had been determined. Phillips emphasized in his *Guide to Geology* that contemporary geology could not yet fix a precise date for the age of the earth; it did, however, prove 'that a long succession of time elapsed during the construction of the visible crust of the globe.'[5]

The common belief that awareness of a long span of geological time was forced on Buckland and his circle by the Huttonians,

[1] *Époques de la nature,* 1778, septième époque.

[2] *Géologie élémentaire,* 1833, 7.

[3] See Burchfield, *Lord Kelvin and the Age of the Earth,* 1975, *passim.*

[4] 'On the Astronomical Causes which May Influence Geological Phaenomena', *Trans. Geol. Soc.* iii (1835), 298.

[5] 1834, 47.

particularly by Lyell,[6] is entirely erroneous. Buckland's work in stratigraphy and paleontology provided concrete evidence for a long time-scale of earth history and produced some of the finest and most widely cited examples of the immensity of geological time. The deposition of strata and formations to a cumulative thickness of several miles had obviously required a very long time. Buckland commented: 'The first impression excited in our minds by the geological examination of any extensive tract of country, is that of the long duration of time which has been occupied in the deposition of the strata that compose the surface of the earth.'[7]

Detailed evidence of the lapse of time was to be found in clasts reworked from older formations and deposited in younger ones, and also in organic accumulations produced by a succession of life cycles. In an early paper to the Geological Society (1816) Buckland described the occurrence of Chalk pebbles in the Plastic Clay, analogous to the *argile plastique* of the Paris Basin, below the London Clay and above the Chalk. These pebbles demonstrated a long sequence of events, i.e. deposition and consolidation of the Chalk, its uplift and denudation, and the transport and deposition of the pebbles to become part of the Plastic Clay; innumerable oyster shells attached to the pebbles added to the evidence for a long lapse of time.[8] Similar evidence was provided by cave deposits accumulated during habitation by many generations of hyenas; the Portland dirt bed produced by forest growth; and coprolites which demonstrated that successive generations of ichthyosaurs had lived and died undisturbed by geological upheavals.

Moreover, there existed entire beds and formations of organic accumulations, such as those composed of coral limestone. 'Strata thus loaded with the exuviae of innumerable generations of organic beings, afford strong proof of the lapse of long periods of time, wherein the animals from which they have been derived lived and multiplied and died, at the bottom of seas which once occupied the site of our present continents and islands.'[9] Such evidence opened up a panorama of vast periods of time. The *Magazine of Natural History* wrote about a particular limestone formation:

[6] E.g. Albritton, *Abyss of Time*, 1980, ch. 11. See also Rudwick, 'Lyell on Etna', *Toward a History of Geology*, ed. Schneer, 1969, 288–304.

[7] Bridgwater Treatise notes, OUM, Bu P.

[8] 'Description of a Series of Specimens from the Plastic Clay', *Trans. Geol. Soc.* iv (1817), 277–304.

[9] Buckland, *Geology and Mineralogy*, i (1836), 116.

By means of the fossils in this order of rocks, we are furnished with unanswerable evidence of the antiquity of our globe, and we can form some vague notions of the vast series of years which must have elapsed during the formation of such a multitude of deposits, and even of the subordinate parts of any one formation; for instance, those of the London Clay, or calcaire grossier. These were evidently deposited slowly, and in a tranquil sea, since the fossils are found in regular beds, and in perfect preservation. It also appears that, after some species were deposited, they wholly disappeared, and gave place to others. All these facts indicate a long series of generations of marine animals.[10]

Further evidence came from the new subject of micropaleontology. Studies of microscopic organisms, pioneered by Christian Gottfried Ehrenberg in Germany, led to the discovery of the volumetric importance of their shells in limestone formations. In a memoir 'On the Calcareous and Siliceous Microscopic Animals which Form the Chief Component Parts of Cretaceous Rocks'(1839), Ehrenberg demonstrated the preponderance of microscopic skeletal material in the Chalk. He calculated that a single cubic inch might contain more than a million tiny skeletons.[11] In a widely reported lecture to the Ashmolean Society 'On the Agency of Animalcules in the Formation of Limestone' (1841) Buckland hailed Ehrenberg's discovery as 'a new and important era in paleontology'. With the aid of thin sections of Stonesfield slate, he illustrated the micro-fossiliferous composition of calcareous rocks.[12] Mantell also was inspired to write about the new subject in two charming booklets, *Thoughts on a Pebble* (1836) and *Thoughts on Animalcules* (1846).

The paleontological evidence for the immensity of time fired the imagination, and it was used by Hack in her *Geological Sketches* (1832) for the young. Harry says to Mrs Beaufoy: 'I cannot imagine how the gradual formation of land by depositions from water, and the tedious labours of coral-worms, is to be reconciled with the account given by Moses. *Six days*! Why six thousand days would not have done it.'[13]

This quotation reflects the context in which the age of the earth · was discussed: whereas, during the second half of the nineteenth century, controversy about geological time took place against the

[10] iii (1830), 72.
[11] *Edinburgh New Phil. Journ.* xxviii (1840), 163.
[12] *Abstracts Proc. Ashmolean Soc.* i (1844), no. xvii, 35–9. See also his *Geology and Mineralogy*, 2nd edn. 1837, i. 610–11.
[13] 38–9.

background of Darwinian evolution and its need for vast lapses of time, during the first half of the century the discussion was bound up with a new interpretation of the first chapter of Genesis intended to reconcile it with geology. Buckland concluded from his exegesis of the first verse of Genesis that 'millions of millions of years may have occupied the indefinite interval, between the beginning in which God created the heaven and the earth, and the evening or commencement of the first day of the Mosaic narrative.'[14]

Buckland's demand for geological time not only upset the biblical literalists, but even irked some of the Huttonians, who regarded a vast lapse of time as their scientific property. In his review of Buckland's Bridgewater Treatise for the *Edinburgh Review* Brewster complained of 'immodesty': 'we must say that if geologists conceive that they add dignity to their science by rash expressions of *millions of millions* of years, they mistake the feelings as well as the judgment of the public'.[15]

An early book entirely devoted to the topic of the *Earth's Antiquity* was written by a country rector, James Gray in 1849; he argued that the great antiquity of the world was in harmony with the Mosaic record of creation. Buckland used this book for his last lecture course at Oxford in 1849, something which Gray proudly acknowledged in his second edition of 1851.

[14] *Geology and Mineralogy*, i. 21–2.
[15] lxv (1837), 13.

# 14
# Organic Evolution Denounced

In the secondary literature there is much confusion concerning the meaning of progressive development. Its acceptance tends to be interpreted as indicating evolutionary leanings.[1] To most English in the 1820s and 1830s progressivism had no connection whatsoever with the idea of organic evolution. The hypothesis of the transmutation of species was known, especially in the context of reptilian paleontology, but it was denounced; till the 1840s evolution did not even constitute a determinant of theory or allegiance in English geology. The English school shared with the early evolutionists a belief in organic continuity and fossil progress, but used these phenomena to argue for creation. It shared with the Scottish Evangelicals a belief in creation, although several Scots tended to oppose the notions of continuity and progressive succession.

Throughout the eighteenth century the 'law of continuity' had been a popular notion; it interpreted nature in its inorganic and its organic aspects as a complete continuum.[2] To scientists of the early nineteenth century this law was associated with the names of Boscovich and Leibniz. It was commonly expressed by means of Linnaeus's dictum, 'Natura, opifex rerum, saltus non facit.' In natural history the law of continuity took the form of the notion of a single linear chain of being, which linked minerals to plants, plants to animals, animals to man, and possibly man to higher beings.

The law of continuity formed a theoretical framework which was favourable to the eighteenth- and early nineteenth-century preoccupation with the classification of minerals, plants, and animals, and also of strata and geological formations. It supported a variety of opposing philosophical and religious beliefs. Continuity was used in natural theology to argue that the chain of being in its fullness indicates creative perfection. Early evolutionists regarded the chain

---

[1] E.g. Corsi, 'The Importance of French Transformist Ideas for the Second Volume of Lyell's *Principles of Geology*', *Brit. Journ. Hist. Sci.* xi (1978), 224.

[2] See Lovejoy, *Great Chain of Being*, 1936.

of being as a temporal continuity produced by the origin of life from inorganic matter and the transmutation of species.[3]

However, the notion of a single linear chain of being came to seem increasingly simplistic, and it became apparent that fundamental breaks exist. Cuvier argued in his *Règne animal* (1817) that the animal kingdom is broken up into four disconnected groups, (1) the radiates, (2) the articulates, (3) the molluscs, and (4) the vertebrates. Continuity could be found within these groups, but not between them. This addition of breaks to continuity did not affect the argument for creative design; but it did present an obstacle to the notion of organic evolution. Some naturalists tried to bridge the gap between molluscs and vertebrates by suggesting that cephalopods constitute an intermediary form. Among those who sympathized with this idea was Geoffroy Saint-Hilaire, who in 1830 engaged in a controversy with Cuvier about this and related matters of taxonomy.[4]

The grand men of early nineteenth-century classification were predominantly French, e.g. Haüy (mineralogy), Jussieu (botany), Lamarck (invertebrates), and Cuvier (vertebrates). However, their taxonomic controversies did not remain confined to Paris but were echoed in disputes among naturalists in Edinburgh and London. For example, a controversy arose in reaction to a paper by James Bicheno 'On Systems and Methods in Natural History' (1827).[5] In England William MacLeay reacted in the *Zoological Journal* (1829), arguing for the law of continuity.[6] In Scotland Fleming wrote a review of Bicheno's paper for the *Quarterly Review* (1829) in which he took the opportunity to criticize the idea of continuity used by both MacLeay and Lamarck and in particular to repudiate the latter's belief in the transmutation of species.[7] Lamarck's theory of organic evolution was more elaborately discussed by Lyell in the second volume of his *Principles* (1832) in which he argued for the fixity of species, in line with Fleming's position.

During the 1820s the debate about the law of continuity in natural history shifted its focus away from the taxonomy of contemporary life towards paleontology and the succession of fossils.

[3] Ibid., ch. 9.
[4] See Bourdier, 'Geoffrey Saint-Hilaire versus Cuvier', *Toward a History of Geology*, ed. Schneer, 1968, 36–61.
[5] *Phil. Mag.* iii (1828), 213–19, 265–71.
[6] 'A Letter to Bicheno', *Zool. Journ.* iv (1829), 401–15.
[7] xli (1829), 302–27.

The study of fossils in their historical sequence became a new testing ground for the relative merits of the various hypotheses about the origin of life. Initially extinction was regarded by some naturalists as incompatible with a plenitude of forms. Each species was believed to represent a necessary link in the chain of being, an integral part of creation as a whole, contributing to its perfection. Destruction of a single link would lead to the dissolution of the entire chain. Divine providence would not let this happen, a belief theologically supported by the story of Noah's Ark which had served to preserve representatives of all species. Even though this philosophy found its greatest advocates (and opponents) outside England, it was represented here as well. Thomas Pennant, the traveller and naturalist, stated: 'Providence maintains and continues every created species; and we have as much assurance, that no race of animals will anymore cease while the earth remaineth, than *seed time and harvest, cold and heat, summer and winter, day and night.*'[8]

The eighteenth-century language of providence and of the chain of being was gradually adjusted to the discoveries of geology. Parkinson stated in the third volume of his *Organic Remains of a Former World* that extinction was part of divine superintendence of earth history. Others argued that fossils are missing links which, added to the array of living forms, fill gaps and produce a more complete chain of being. The significane of fossils as links in the chain of being was used by John Miller in a prospectus for his *Natural History of the Crinoeidea* (1822); because of geological discoveries 'many very important links in the scale of organized Nature, which must otherwise have presented an imperfect and broken chain, have been by these means satisfactorily supplied.'[9]

Probably the most notable example of extinction, discussed in the context of a belief in providence and a chain of being, was the dodo, which became extinct around 1690. Its destruction in very recent, historical times linked contemporary zoology to the paleontological sequence of extinctions. The dodo was one of several birds of the Mascarene Islands, of which Mauritius is the best-known, which became extinct when the islands were used as a stop-over on the trade route between Europe and the East Indies. In 1650 a dodo was brought to London and stuffed for exhibition at the

[8]Quoted by White, 'Observations on a Thigh Bone of Uncommon Length', *Mem. Lit. Phil. Soc. Manchester* ii (1785), 351.
[9]1.

Tradescant Museum, but moved in 1683 to the Ashmolean Museum in Oxford. The specimen was subsequently destroyed except for its head and one leg, which are now exhibited in the University Museum in Oxford, together with an original painting of a dodo by John Savery. Some naturalists of the early nineteenth century doubted the former reality of the dodo, but in the *Zoological Journal* for 1828 Duncan proved its historicity.[10] The taxonomic affinity of the dodo, a member of the Columbiformes, remained a source of controversy until the 1840s. All previous literature on the subject was eclipsed by Strickland's classic monograph of 1848, *The Dodo and its Kindred*.[11]

The dodo showed that extinction does occur, and that the destruction of a link in the chain of being does not trigger a domino effect of species annihilation. John Thompson commented in the *Magazine of Natural History* in 1829:

If we seek to find out what link in the chain of nature has been broken by the loss of this species, what others have lost their check, and what others necessarily followed the loss of that animal which alone contributed to their support, I think we may conclude that, the first being foreseen by the Omniscient Creator, at least no injury will be sustained by the rest of the creation; that man, its destroyer, was probably intended to supplant it, as a check; and that the only other animals which its destruction drew with it, were the intestinal worms and Pediculi peculiar to the species.[12]

Buckland placed the dodo at the front of the paleontological sequence of extinctions on the fold-out plate of the crust of the earth published with his Bridgewater Treatise. He contributed to the transference of the language of natural theology from contemporary forms of life to the geological record of extinctions. To Buckland fossils were links 'that appeared deficient in the grand continuous chain which connects all past and present forms of organic life, as part of one great system of Creation'.[13] In this view continuity in the chain of being existed only when one added past species to present ones. Thus plenitude became a historical notion; the chain of being would have no deficiences if considered as a chain of history. All past life should be intercalated in the sequence of

---

[10] 'A Summary Review . . .', *Zool. Journ.* iii (1828), 554–66.

[11] See Hachisuka, *Dodo and Kindred Birds*, 1953.

[12] 'Contributions Towards the Natural History of the Dodo', *Mag. Nat. Hist.* ii (1829), 448. See also Mantell, *Geology of the South-east of England*, 1833, 357–9.

[13] *Geology and Mineralogy*, i (1836), 88.

current species. The fossil pachyderms from the Tertiary of the Paris Basin were interpreted as forms which filled the taxonomic gaps among its living relatives. When the *Sivatherium* was discovered in the Siwalik Hills, Buckland hailed it as the link connecting pachyderms with ruminants. 'This discovery, amid the relics of past creations, of links that seemed wanting in the present system of organic nature, affords to natural Theology an important argument, in proving the unity and universal agency of a common great first cause; since every individual in such an uniform and closely connected series, is thus shown to be an integral part of one grand original design.'[14]

However, the bearing of fossils on the belief in organic evolution was not lost to its sympathizers. Robert Grant, while at Edinburgh, wrote an anonymous piece for the *Edinburgh New Philosophical Journal* (1826), in which he sided with Lamarck on the basis of the phenomenon of fossil progression.[15]

Fleming, in his review of 1829, rejected not only continuity among contemporary forms of life, but also progression among fossils, either in support of divine design or of organic evolution. He believed that the various branches of natural history ought to be studied as unconnected, autonomous realms:

The strata present to the student the relics of various groups of organized beings; but these must be examined in the peculiar compartments which have been allocated to them. The fossils of the chalk rocks must not be mingled with those of the carboniferous limestone, nor with the *species* which now exist. All these must be studied as *separate systems*—the works of the same Omnipotent Creator—formed for particular purposes, and existing during different epochs—'of the capacious plan, which heaven spreads wide before the view of man'.[16]

The zoological aspects of Lamarck's advocacy of organic evolution were discussed by Lyell in 1832. However, the paleontological aspects of evolution had already been a subject of discussion for more than a decade among English geologists, as a consequence of the discovery of fossil reptiles. From Home in 1814 to Owen in 1841 the history of reptilian paleontology charted the attitude of

---

[14] Ibid., 114.
[15] 'Observations on the Nature and Importance of Geology', *Edinburgh New Phil. Journ.* Apr. to Oct. 1826, 297.
[16] *Quarterly Review* xli (1829), 327.

English geologists to the issue of progressive continuity among fossils and its implications for theories of the origin of species.

When Home described the ichthyosaur in 1814 he interpreted it as a missing link in the chain of being.[17] A similar opinion was expressed by Conybeare when he described the plesiosaur in 1821. He too believed that the discovery of fossil reptiles 'adds new links to the connected chain of organized beings'.[18] He explained in some detail what he meant by a continuity in the chain of being, in contrast to the evolutionary interpretation. Conybeare lucidly formulated the ecological criterion of continuity, which during the 1820s became the criterion of progressive development as worked out by Buckland:

When alluding to the regular gradation, and, as it were, the linked and concatenated series of animal forms, we would wish carefully to guard against the absurd and extravagent application which has sometimes been made of this notion. In the original formation of animated beings, the plan evidently to be traced throughout is this. That every place capable of supporting animal life should be so filled, and that every possible mode of sustenance should be taken advantage of; hence every possible variety of structure became necessary, many of them such as to involve a total change of parts, but others again, such as required nothing beyond a modification of similar parts, slight indeed in external appearance, yet important in subserving the peculiar habits and economy of the different animals; in these cases the unity of general design was preserved, while the requisite pecularity of organisation was superinduced; nor can there be any where found a more striking proof of the infinite riches of creative design, or of the infinite wisdom which guided their application.[19]

Conybeare ended his definition of organic continuity with an account of Lamarck's belief in the transformation of species, which he denounced as so monstrous 'that nothing less than the credulity of a material philosophy could have been brought for a single moment to entertain it—nothing less than its bigotry to defend it'.[20]

However, the evolutionary hypothesis could not be brushed aside so easily, especially after Saint-Hilaire had joined the fray. Among the discoveries of fossil reptiles during the period of 1814–24 was that of the teleosaur, an extinct crocodilian form similar to the

[17] *Phil. Trans.* civ (1814), 572.
[18] *Trans. Geol. Soc.* v (1821), 560.
[19] Ibid., 560–1.
[20] Ibid., 561.

modern gavial. Teleosaurian fossils were known from Caen in France and a very fine specimen was discovered in the Lias of Whitby in 1824.[21] Saint-Hilaire speculated that the various reptilian forms (ichthyosaur, plesiosaur, pterodactyl, teleosaur) constituted a temporal, genetic sequence, each form having given rise to the next and the teleosaur to such mammals as the megatherium. Around 1830 Saint-Hilaire read a number of memoirs to the Académie des Sciences, published as his *Recherches sur de grands sauriens trouvés à l'état fossile* (1831), in which he presented his reptilian research and outlined a hypothesis of evolution by transmutation of species caused by environmental changes, especially in the composition of the atmosphere.[22]

Saint-Hilaire's evolutionary hypothesis came to the attention of English geologists not only through the *Edinburgh New Philosophical Journal* but also by way of copies of his memoirs sent to colleagues. Buckland, who was among the recipients, studied Saint-Hilaire's papers carefully and began a file which, though entitled 'Species change of Lamarck', dealt primarily with evolutionary ideas based on reptilian paleontology. The file was part of the preparations for his Bridgewater Treatise, but the published version mentioned only disapproval of Lamarck's hypothesis and did not allude to Saint-Hilaire's by name.

The file gives a clear picture of Buckland's views. To him the idea of species change was absurd because it was non-empirical and non-actualistic; he scribbled: 'In geological investigations we find a repeated succession of creations. In the natural world nothing like creation now proceeding. All that we witness is the derivation of new individuals by the process of parentage from preceding individuals of the same species.'[23] The file also contains sarcastic remarks directed at the crocodile–man relationship, betraying a concern with the implications of evolution for man's origin. One sentence reads: 'I am not quarreling or finding fault with a crocodile; a crocodile is a very respectable person in his way, but I quarrel with the calling a man a *crocodile improved*.'[24] Elsewhere he says: 'We shall not be degraded from our high estate to say to that reptile crocodile: "Trust right loving cousin we greet you well".'[25]

[21] Young, 'Account of a Fossil Crocodile', *Edinburgh Phil. Journ.* xiii (1825), 76–81.
[22] See n. 4.
[23] OUM, Bu P.
[24] Ibid.
[25] Ibid.

Some of the file's more substantial arguments did end up in the Bridgewater Treatise. Buckland pointed out that the various reptilian forms of Saint-Hilaire's reptilian lineage do not originate in successive geological formations, but simultaneously at an early stage as depicted in *Duria Antiquior.* He concluded that such contemporaneity precludes descent 'by any process of gradual transmutation or development'.[26]

Buckland extended this line of argument to the fossil record *in toto;* he observed that representatives of the four groups into which Cuvier had divided the animal kingdom occur simultaneously in the oldest fossiliferous deposits. Thus these groups could not have descended one from the other but must have run parallel in the course of earth history. Within each group the anatomical ground-plan had remained unchanged, in spite of many successive variations on it. These had been perfectly adapted to changes in the environment, proving creative design, and being utterly inexplicable by the evolutionary mechanisms proposed by Lamarck or Saint-Hilaire, which ought to have produced imperfections and monstrosities.

Grant, who moved to London to become professor of zoology at the new university, kept up his support for Lamarck and advocated the evolution hypothesis openly in his lectures published in the *Lancet* (1834). Grant was quoted by Owen in his 'Report on British Fossil Reptiles' (1841), where he argued vigorously against the hypothesis of the transmutation of species. To Buckland's argument against reptilian evolution Owen added that the reptiles had originated suddenly, showing a wide gap of organization with the antecedent fishes.

There was, nevertheless, something peculiarly alluring in the idea of organic evolution seen in the light of historical geology. This became apparent when Chambers published his *Vestiges of the Natural History of Creation* (1844, anonymously) which went through four editions in just half a year.[27] The bulk of the *Vestiges* was little more than a summary of contemporary historical geology, though it also dealt with taxonomy, human society, and the mental condition of animals. Chambers appropriated the very core of historical geology to argue that organic evolution had occurred. His line of argument was as follows. Astronomers believe in the nebular hypothesis of planetary origin; geologists believe in the thermal

---

[26] *Geology and Mineralogy,* i (1836), 254.
[27] See Millhauser, *Just Before Darwin,* 1959, *passim.*

evolution of the earth as a planet; paleontologists have shown how life has originated in successively progressive stages and how extinctions have occurred by such natural means as climatic change. Why then should one not believe that species have originated and been transformed by natural means as well, by laws of nature, and so restrict the evidence for design to the laws themselves rather than extend it to their innumerable products? This line of argument added very little, if any, new material, but pillaged the armoury of the English school and used its own weapons against it.

Both Sedgwick in 1831 and Whewell in 1837 had objected to the transmutation of species. Both were invited to write a review of the *Vestiges* for the *Edinburgh Review,* and Sedgwick did so (1845).[28] It turned out to be a scathing attack of no less than eighty-five pages. Sedgwick's reaction was not only one of scientific criticism or orthodox disbelief, but of moral revulsion; a hypothesis which attributed the origin of species, and of man himself, to natural causes seemed a case of crass materialism if not moral depravity, imported from abroad, possibly from Scotland, but alien to England; a target more for scorn and ridicule than for serious consideration. Sedgwick reacted to Chambers's idea as if it were a sexual perversion: it was unnatural, its lure to be suppressed by violent denunciations, and unmentionable in the presence of ladies. Sedgwick fumed:

If our glorious maidens and matrons may not soil their fingers with the dirty knife of the anatomist, neither may they poison the springs of joyous thought and modest feeling, by listening to the seductions of this author; who comes before them with a bright, polished, and many-coloured surface, and the serpent coils of a false philosophy, and asks them again to stretch out their hands and pluck forbidden fruit—to talk familiarly with him of things which cannot be so much as named without raising a blush upon a modest cheek;—who tells them—that their Bible is a fable when it teaches them that they were made in the image of God—that they are the children of apes and the breeders of monsters—that he has *annulled all distinction between physical and moral,*—and that all the phenomena of the universe, dead and living, are to be put before the mind in a new jargon, and as the progression and development of a rank, unbending, and degrading materialism.[29]

---

[28] See Brooke, 'Richard Owen, William Whewell, and the *Vestiges'*, *Brit. Journ. Hist. Sci.* x (1977), 132–45.

[29] lxxxii (1845), 3.

Sedgwick's review, while being full of invective and criticism of
scientific detail, initiated a new emphasis on discontinuity in earth
history, thereby backing away from his earlier position, which had
stressed continuous progressive change. He had explicitly supported
the nebular hypothesis and its corollary of central heat in his
*Discourse;* and in an anniversary address to the Geological Society he
had advocated progressive development: 'I think that in the repeated
and almost entire changes of organic types in the successive formations of the earth . . . we have a series of proofs the most emphatic
and convincing . . . that the approach to the present system of things
has been gradual, and that there has been a progressive
development of organic structure subservient to the purposes of
life.'[30] This stress on continuous historical progress had been
intended as a refutation of Lyell; but its use by Chambers caused
Sedgwick to back-track and shift his emphasis. In his review he
called the nebular hypothesis no more than 'a splendid vision', and,
instead of gradual progress, he underscored discontinuity in the
paleontological record: types appear suddenly, highly organized,
remain for a long time without much change, and equally suddenly
become extinct; linking gradations between different groups are
absent; in some instances degeneration rather than progressive
development is evident.[31] Sedgwick enlarged his review and added
it as a lengthy introduction to the fifth edition of his *Discourse* (1850).

This change of emphasis, and the strong feelings which were
generated by the organic evolution debate obscured old intellectual
alliances and contributed to new ones. In contrast to the disapproval
directed at the first volume of Lyell's *Principles,* the second volume
was approvingly cited. Buckland, who had been severely criticized
in the past by such Scottish Evangelicals as Brewster or Fleming,
now found himself in alliance with them against the evolutionary
philosophy. One Scottish Evangelical in particular, Hugh Miller,
extended his influence south of the border. Buckland had already
lauded his work on the Old Red Sandstone and on fossil fish in the
early 1840s; but in the late 1840s he also made use of Miller's
arguments against Chambers. In 1849 his course announcement
read: 'The Reader in Geology will begin a Course of Lectures . . .

---

[30] *Proc. Geol. Soc.* i (1834), 307–8.
[31] See Sedgwick's correspondence with Agassiz about evolution, E. C. Agassiz, *Life
and Correspondence of Louis Agassiz,* i (1885), 383–97.

demonstrating by Evidences in the Museum . . . the facts cited in . . . Hugh Millers's ''Footsteps of Creation,'' shewing the Fallacies of the Doctrine of *Development* maintained by the anonymous Author of the ''Vestiges of Creation.'' '[32]

[32] OUM, Bu P.

# 15
# The English School of Historical Geology

When the geology lecture courses at Oxford and Cambridge began, in the 1810s, few of the famous names in the subject were English. But during the 1820s the English contributions to the study of earth history began to rank among the finest and most sensational internationally. Detailed descriptions of Secondary formations, revolutionary additions to paleontology, together with a host of other innovative contributions, showed that English geology was drawing level with that of Germany and France, and, in its own estimation, pulling ahead of that of Scotland.

The preceding decades had given to geology a vastly expanded data base, acquired largely by inductive work carried out under the Baconian motto of the Geological Society. Geology, it seemed, had definitively exorcized the evil spirit of idle speculation and of deductive cosmogony, and had risen to the rank of an inductive science. The geologists at Oxford and Cambridge were now counted among the famous scientists in the country, and within the two universities attendance at their lecture courses had become a symbol of aspiration to academic excellence. The first volume of Loudon's *Magazine of Natural History* (1829) exclaimed in a review of one of Buckland's papers: 'When we look at the state of English geology now, ennobled by the collateral sciences, and almost essential to a liberal education, we are led to forget that it is a science of our own times, that most of its earliest professors are yet amongst us.' [1]

This rapid advance of English geology led by about 1830 to a need for stock-taking. The extensive base of descriptive knowledge now warranted, if not demanded, a fresh attempt at theory and synthesis. Accordingly, a number of review papers appeared in which prominent members of the English school summed up its accomplishments, contrasted them with those of other nations and

---

[1] i (1829), 249.

schools of geology, defined its theoretical position, chose its intellectual ancestry, and contemptuously dismissed mutterings, being heard just then, about a 'decline of science in England'.

The self-definition of the English school centred, as far as its theory of geology was concerned, on geology as a form of history. The most sensational contribution to earth history during the 1820s had been the diluvial theory. However, diluvialism was now eclipsed by a new and wider-ranging synthesis of historical geology, that of geological progressivism. Nothing gave a better indication of the scientific quality and the grandeur of this new perspective of earth history than the famous accolade awarded to geology by Herschel in his *Preliminary Discourse on the Study of Natural Philosophy* (1830): 'Geology, in the magnitude and sublimity of the objects of which it treats, undoubtedly ranks, in the scale of the sciences, next to astronomy.'[2] Geology, as earth history, and with the aid of its inductive philosophy, had climbed in little more than two decades from the pedestrian level of natural history to the very top of the scientific hierarchy. No quotation was used more often during the 1830s than this sentence by Herschel, and it was paraphrased in a number of ways.[3]

The most explicit self-definition was formulated by the two 'party theoreticians' of English geology, Conybeare and Whewell. The latter, especially, had a talent for lucid generalization which he used to outline the history of geology and the place of geology among the sciences in the English universities. In a review of the progress of geology Whewell sweepingly divided the subject into four schools corresponding to 'the four principal scientific nations of Europe'. *Primary* geology was a product of Germany where Werner and his followers had worked out the lowest part of the geological succession and had concentrated on its mineralogical properties. France could lay claim to the geology of the *Tertiary*, the upper part of the succession, elucidated by Cuvier and his collaborators. The most recent, or *Quaternary* deposits, had been successfully studied in Italy, in particular by Lyell who by this time was no longer regarded as a member of the English school.

However, the extensive middle part of the geological record, the

[2] 287.
[3] Buckland: 'The history of the earth next to that of the heavens affords the most sublime subject of study that can be derived from physical science.' Bridgewater Treatise notes, OUM, Bu P.

*Secondary*, 'belongs in a very considerable measure to England'. England had become 'the head-quarters of secondary geology', an island blessed with a uniquely condensed and yet distinct series of most fossiliferous rocks.[4] As a result English stratigraphy had become the example for the rest of Europe, the standard to which corresponding strata on the Continent were referred. This success in the study of stratigraphy 'must make it hereafter appear one of the most remarkable passages in the history of science'.[5] Whewell grew lyrical in his description of the English school:

> The English school of geology has long consisted of a body of active, inquiring men, eager suitors of truth, forming their opinions principally by the use of their own eyes, feet, and hands; by the intercourse of conversation; and by a diligent perusal of each other's writings, which, from various causes, were little studied and not generally known beyond their own circle. Their disciples have been taught, their converts and teachers formed, with the hammer in their hands, with the knapsack at their backs, or at their saddlebow, in long and laborious journeys, amid privation and difficulty; with a perpetual personal exercise, both of the most minute discrimination of differences, and of the widest sweep of combination.[6]

Conybeare sounded no less assertive about the English accomplishments in his *Report* on the state of geology. He emphasized that the study of fossils rather than of minerals had been a root cause of the success of English geology and that the 'English school has distinguished itself by the ardent and successful zeal with which it has developed the whole of the secondary series of formations: in these the zoological features of the organic remains associated in the several strata, afford characters far more interesting in themselves and important in the conclusions to which they lead, than the mineral contents of the primitive series.'[7]

This summary of the contributions by the English school was more than an exercise in the history and philosophy of science; it was an expression of chauvinism and of pride in the participation of geology in national progress, reform, and expansion of the Empire. Conybeare ended his *Report* by emphasizing the desirability of broadening the scope of geological research, to include not only America, but also Australia and India: 'England, the mistress of

[4] 'Progress of Geology', *Edinburgh New Phil. Journ.* Apr. to Oct. 1831, 242–7.
[5] 'Lyell—*Principles of Geology*', *British Critic* ix (1831), 180.
[6] Ibid., 181.
[7] *Report BAAS, 1831, 1832*, 370.

such vast and remote portions of the globe, seems peculiarly called upon to take the lead in this task. And the increased attention to scientific pursuits, now diffusing itself among her military and naval classes,—one of the most favourable characteristics of the age,—promises to supply her every day with observers more and more competent to achieve this honourable duty.'[8]

Conybeare gave expression on the newly acquired status of geology by choosing as its forerunner none less than Leibniz, the most eminent and admired theist philosopher of the modern age. He professed to see in Leibniz's *Protogea* the beginning of a truly historical form of geology, the notion of a central igneous agency, a programme for the historical investigation of geological phenomena, and a prudent empiricism. Leibniz, moreover, had felt that scholarship was an area where men of different denominations could co-operate harmoniously, as did the British Association to which Conybeare read his *Report*.[9]

Sedgwick in his *Discourse* and Buckland in his Bridgewater Treatise gave further form and definition to the English school, in particular to its synthesis of geological progressivism. Buckland developed in some detail the theist notion that earth history has a distinct beginning in time. 'Geology has already proved by physical evidence, that the surface of the globe has not existed in its actual state from eternity, but has advanced through a series of creative operations, succeeding one another at long and definite intervals of time.'[10] This perspective of the past was at variance with the deist notion that history is a cyclical process of recurrences to which one can attribute neither a beginning nor an end. But the new geology attributed a beginning not only to the present order of things, but to a series of antecedent creations as well, thus invalidating the Huttonian theory of the earth:

Against this theory, no decisive evidence has been accessible, until the modern discoveries of Geology had established two conclusions of the highest value in relation to this long-disputed question: the first proving, that existing species have had a beginning, and this at a period comparatively recent in the physical history of the globe: The second showing that they were preceded by several other systems of animal and vegetable life,

[8] Ibid., 413.

[9] Davy had already singled out Leibniz. See Siegfried, 'Davy's Lectures at the Royal Institution', *Science and the Sons of Genius*, ed. Forgan, 1980, 187.

[10] *Geology and Mineralogy*, i (1836), 10–11.

respecting each of which it may be no less proved, that there was a time when their existence had not commenced; and that to these more ancient systems also, the doctrine of eternal succession, both retrospective and prospective, is equally inapplicable.[11]

Buckland carried this one step further, arguing that it was a priori implausible if not impossible that any organisms could have existed during the hot, early stages of earth history. 'In these most ancient conditions, both of land and water, Geology refers us to a state of things incompatible with the existence of animal and vegetable life; and thus, on the evidence of natural phenomena, establishes the important fact that we find a starting point, on this side of which all forms, both of animal and vegetable beings, must have had a beginning.'[12] During the primeval incandescance of the earth, in particular, nothing organic could possibly have existed.

The various textbooks and encyclopaedia entries written by de la Beche or by Phillips during the 1830s further expounded the position of the English school. Towards the end of the decade new heights of accomplishment were reached when Murchison's *Silurian System* (1839) finally appeared. It represented a significant extension of the stratigraphical territory on which the English school had founded its reputation, adding to the Secondary succession the considerable extent of the Transition rocks, previously conceded to Germany and the Wernerians. Murchison's work was a model specimen of professionalism, long on data and short on speculation. It provided new support for the progressivist synthesis, the perspective according to which earth history was driven by central heat, channelled through progressive succession, and punctuated by global upheavals.

On the fringe of the English school itinerant lecturers and their textbooks disseminated its discoveries and theoretical conclusions. W. Mullinger Higgins or Robert Bakewell were prominent examples of provincial proselytes. The latter wrote to Silliman shortly after the appearance of a new edition of his *Introduction to Geology* (1833) that Buckland and Sedgwick had sent him congratulations and approbation. 'Professor Buckland also told a gentleman whom I knew, that it was decidedly the book he should choose to place in the hands of his pupils.'[13]

[11] Ibid., 54.
[12] Ibid., 53.
[13] 24 Sept. 1833, Fisher, *Life of Silliman,* ii (1866), 54.

## Contrast with Scottish Geology

When the Oxford school began to take shape in the course of the 1810s it did so in self-defining contrast to the Edinburgh school with its Huttonian and Wernerian protagonists. But during the 1820s, when Oxford geology was included under the English school, the contrast with the majority of the Wernerians diminished. At the same time tensions with the Huttonian Scots remained and came to a head when Lyell published his *Principles*, criticism of which by the English school served to emphasize its distinct identity.

Opposition of a modified Wernerian kind to the English interest in Secondary paleontology and stratigraphy was put up by MacCulloch. He had been one of the most active early members of the Geological Society, and his papers on crystalline rocks had culminated in his *Geological Classification of Rocks* (1821). A decade later he published a new *System of Geology* (1831) in which he scorned the paleontological discoveries of the preceding decade. MacCulloch believed that no fundamental fact had been added to geology during the 1820s. If Herschel's accolade was the most famous quote of the time, MacCulloch's pronouncement about geological discoveries was the most infamous, namely: 'I do not perceive that a new one has been added to the Science.'[14]

This was the stubborn stance of a man who had been unable to move with the times and who resented the way in which the glamour of paleontology had overshadowed the study of mineralogy, which had been the basis of his early prominence. Reviewers of MacCulloch's *System* were harsh in their criticism of his book; the censorious style and reactionary stance seemed to invite sarcasm. Murchison, in his anniversary address to the Geological Society of 1832, combined disapproval of the book with clarifications of the position of the English school. He summed up: 'This work, in short, is so far from being a new system, that it can hardly be said to enter into the boundless field now opened to modern geologists—the evidence derived from organic remains, the very key-stone of our fabric, being either slightly touched upon, or its value derided.' MacCulloch's book represented spiteful obscurantism: 'It is indeed by the help of zoological distinctions that modern geology has been carried onwards far beyond the original

---

[14] *System of Geology*, i (1831), v. See also Cumming, 'John MacCulloch', *Notes and Records Roy. Soc.* xxxiv (1980), 155–83.

scope of certain earlier observers, who now seem to feel regret that
they can no longer confine it within those mineralogical barriers
with which they had endeavoured to surround it.'[15]

More serious opposition, especially to the progressivist synthesis
of the English school, came from Lyell. The first volume of his
*Principles of Geology* presented a revived and fully fledged Huttonian
theory of the earth. To Hutton, geological change had been cyclical
and indefinite rather than progressive and finite; he had argued that
the earth is in a steady state, because despite change a stable balance
of decay and regeneration is maintained. This belief had been
expressed in Hutton's famous maxim at the conclusion of his
'Theory of the Earth' (1785): 'The result, therefore, of our present
inquiry is, that we find no vestige of a beginning,—no prospect of
an end.'[16]

This was a fundamentally anti-historical view of the past, readily
identified with the eighteenth-century philosophy of deist eternalism and historical agnosticism. But Lyell resuscitated the
Huttonian system with new ideas and a plethora of novel observations. He expressed a deep belief in, if not commitment to, an
absolute uniformity in nature, an invariable constancy in both the
organic and the inanimate world. This system of the earth, illustrated by the anology of constant and circular movement of planets
around the sun, was fundamentally incompatible with any perspective of linear and progressive change. Accordingly, Lyell rejected
the theory of a central heat with its corollary of secular cooling; more
daringly, he objected to the notion of fossil progress. Among his
objections were: (1) in the Primary strata the fossil record may be
partially obliterated and incomplete; (2) the known Transition
rocks were deposited in a marine environment, so we cannot know
what terrestrial animals existed at this time; (3) in the Secondary
rocks there are exceptions to the rule of progress, such as the
mammalian didelphys. Lyell also pointed out complexities in the
Tertiary series and concluded: 'It is, therefore, clear, that there is no
foundation in geological facts, for the popular theory of the successive development of the animal and vegetable world, from the
simplest to the most perfect forms.'[17]

[15] 17 Feb. 1832, *Proc. Geol. Soc.* i (1834), 376–7. See also *Literary Gazette,* 3 Sept. 1831,
561–3. Even Brewster's *Edinburgh Journ. Sci.* pronounced a negative verdict, v (1831),
358–75.

[16] *Trans. Roy. Soc. Edinburgh* i (1788), 304.

[17] *Principles of Geology,* i (1830), 153.

Lyell's belief in absolute uniformity required that a recent class like the mammals, known mainly from the Tertiary, had been in existence as early as the Primary, and that such ancient and extinct monsters as the Liassic reptiles might still live in some unexplored region of the globe. The respective domination of these vertebrate classes during particular geological periods had been a function of the 'season' of a 'great year' of climatological change rather than an irreversible trend of paleontological succession. The season of the reptiles, for example, would some time in the future recur: 'Then might those genera of animals return, of which the memorials are preserved in the ancient rocks of our continents. The huge iguanodon might reappear in the woods, and the ichthyosaur in the sea, while the pterodactyle might flit again through umbrageous groves of tree-ferns.'[18] Somewhat inconsistently Lyell believed that man himself was of very recent origin.

Lyell included in his *Principles* a stratigraphic table, the bottom end of which he left open, omitting any indication that the history of the earth had a recognizable beginning. In his anniversary address of 1837 Lyell reviewed Buckland's Bridgewater Treatise, taking issue with his former teacher's belief in a definite beginning to earth history. Lyell believed that it was wrong even to look for traces of such a beginning. 'Already has the beginning of things receded before our researches to times immeasurably distant. Why then, after wandering back in imagination through a boundless lapse of years, should we expect to find any resting-place for our thoughts, or hope to assign a limit to the periods of past time . . . ?'[19]

The Huttonians had some of the wind taken out of their sails by the English school. The latter had made two tenets of Hutton's theory of the earth integral parts of the progressivist synthesis: that of igneous activity incorporated in the hypothesis of a central heat and that of a very long (though not indefinite) geological time. Any debt acknowledged by the English for these ideas was to continental geologists, not to the Scots. Brewster's review of Buckland's Bridgewater Treatise for the *Edinburgh Review* displayed annoyance

[18] Ibid., 123. See also Fleming's criticism of climatic change, *Edinburgh New Phil. Journ.*, Oct. 1828 to Mar. 1829, 277–86. Conybeare wrote a rejoinder, ibid., Apr. to Oct. 1829, 142–52.

[19] 17 Feb. 1837. *Proc. Geol. Soc.* ii (1838), 522–3. See also Rudwick, 'Uniformity and Progression', *Perspectives in the History of Science and Technology*, ed. Roller, 1971, 209–27; Ospovat, 'Lyell's Theory of Climate', *Journ. Hist. Biol.* x (1977), 317–39; Lawrence, 'Lyell versus the Theory of Central Heat', ibid., xi (1978), 101–28.

with the arrogance of the English. Its introductory part was a defence of the Huttonians and the Scottish tradition. The Scots, Brewster believed, had begun modern geology: 'Professor Playfair and Sir James Hall were the main supporters of the new philosophy of the globe; and if we regard Cuvier as its Newton, Dr. Hutton will be its Copernicus; and the honours of Kepler and Galileo will fall to the lot of Playfair and Hall.'[20] The subsequent work by the English had simply served as the 'vindication of our illustrious countrymen'. Brewster noted with some bitterness the clerical patronage of the English school: 'After having for half a century "stumbled on the dark mountains"', the Church is now feeding her flock on the green pastures of the Huttonian geology.'[21]

This was more Scottish chauvinism than objective history of geology. The progressivist synthesis was at variance with the most basic tenet of Huttonianism, its idea of cyclical indefiniteness. This was recognized by Lyell, who made progressivism the main issue.

The Huttonians may have been justly annoyed at the unacknow-ledged inclusion of some Scottish ideas in the progressivist synthesis, but Lyell's book was not much of a model either in acknowledging debt or in portraying fairly the position of his opponents. The English school was particularly irritated by Lyell's implicit misrepresentation of their philosophy of geology. He made it seem that the Huttonians were unique in interpreting the past by laws now in operation, in contrast to Buckland and his school who allegedly invoked, if not miraculous events of divine intervention, at least causes different in kind and degree from those operating at present.

This was untrue and confused the issue, as reviewers of Lyell's work emphasized. No member of the English school disputed that the laws of nature had been anything but constant. The laws of organic life in particular had been assumed invariable, the very basis of the progressivist synthesis and of the paleontological work of Buckland, Cuvier, and others. The issue was not the constancy of natural laws, but the constancy of their force. Buckland wrote: 'The same causes in all periods of the globe appear to have produced the same effects, but the circumstances under which they operated and the intensity of their force may have been very different.'[22]

---

[20] lxv (1837), 10.
[21] Ibid., 14.
[22] File 'Mechanism of Geology', OUM, Bu P.

Lyell went much further. He rejected this position as the barren descriptiveness of an over-cautious generation, and he asserted, a priori, that the level of intensity of geological processes in the past had not been any different from that of the present. Lyell used the rule that the present is the key to the past as a procrustean device; any evidence weighing against absolute uniformity was either ignored or forcibly explained away.

Conybeare immediately protested at this doctrinaire approach, and in a discussion of Lyell's views for the *Philosophical Magazine* (1830–1) he spelled out precisely what divided them:

I would here for once observe, that the geologists whose cause I advocate, seem from the frequent employment of the phrases *"existing causes"*, and *"the uniformity of Nature"*, to lie under a misconception in Mr L.'s mind; as though they speculated on causes of a different order from any with which we are acquainted, and almost reasoned on the supposition of different laws of nature: Whereas I conceive both parties equally ascribe geological effects to known causes, viz. to the action of water, and of volcanic power; only those with whom I class myself maintain, that much which has resulted from aqueous action, e.g. the excavation of many valleys, indicates rather the violent action of mighty diluvial currents, than effects which do or can result from the present drainage, by the actual rivers, of the waters descending from the atmosphere in rain, etc., to which Mr Lyell looks exclusively.[23]

Sedgwick joined the chorus of protest and took Lyell to task in his anniversary address of 1831. Habitually inclined to express himself in a forceful manner, Sedgwick accused Lyell of ignoring the work which the English school had done in Secondary formations; he saw Lyell's theory as a case of special pleading which reminded him of his training as a lawyer: 'It cannot, I think, be doubted, that in the general statement of his results, Mr Lyell has, unconsciously, been sometimes warped by his hypothesis, and that, in the language of an advocate, he sometimes forgets the character of an historian.'[24] Buckland disapproved of the sharpness of this attack: 'I think it all fair to praise an author in a President's speech recording the events of the year, but hardly right to indulge severe criticism when there is no opportunity of defense or reply.'[25] Buckland preferred the milder

---

[23] 'An Examination of those Phaenomena of Geology, which Seem to Bear most Directly on Theoretical Speculations', *Phil. Mag. Ann. Chem.* viii (1830), 360.

[24] 18 Feb. 1831, *Proc. Geol. Soc.* i (1834), 303.

[25] Buckland to Featherstonhaugh, 28 Feb. 1831, CUL, Se P.

review written by Whewell for the *British Critic*. Whewell expressed his criticism particularly lucidly a few years later in his *History of the Inductive Sciences* (1837):

The effects must themselves teach us the nature and intensity of the causes which have operated; and we are in danger of error, if we seek for slow and shun violent agencies further than the facts naturally direct us, no less than if we were parsimonious of time and prodigial of violence. *Time,* inexhaustible and ever accumulating his efficacy, can undoubtedly do much for the theorist in geology; but *Force,* whose limits we cannot measure, and whose nature we cannot fathom, is also a power never to be slighted: and to call in the one to protect us from the other, is equally presumptuous, to whichever of the two our superstition leans.[26]

This criticism was fully justified. Lyell's position of extreme uniformity was both poor science and bad logic. It is astonishing that later geologists and historians have so consistently and for so long sided with Lyell and dismissed his clerical opponents.

Lyell added to English geology by his extensive and detailed descriptions of contemporary geological processes. Buckland and his school were first and foremost historical geologists. This left vacant the niche of physical and dynamic geology, a vacancy which Lyell filled and in which he was appreciated if not admired by his English colleagues. Buckland, for one, used Lyell's book for his lecture courses, though with a caveat about the doctrine of absolute uniformity. Moreover, Lyell's speculation about climatic change by changes in the distribution of land and sea, intended to replace the theory of a central heat, was regarded as ingenious; after all, to the English school itself central heat was merely a hypothesis. But Lyell's objections to fossil progress were regarded as little more than a form of self-inflicted blindness. Bakewell wrote to Silliman, shortly after the appearance of Lyell's first volume: 'If you have seen it, you will think there is much Scotch amplification. A Scotchman can never write briefly and directly to the point.' Bakewell praised Lyell's observations on recent geology and on climatic change, but considered his objections to progressive development as 'far from satisfactory'.[27]

Lyell's objections to the progressivist synthesis were indeed of low intellectual calibre; they ranked with the obscurantist

[26] iii (1837), 616. See also Hooykaas, *The Principle of Uniformity,* 1963, *passim.* Gould, 'Is Uniformitarianism Necessary?', *Am. Journ. Sci.* cclxiii (1965), 223–8.
[27] 16 Nov. 1830, Fisher, *Life of Silliman,* ii. 53.

objections made by Greenough, MacCulloch, and the biblical literalists, using negative evidence and emphasizing exceptions rather than accepting the rule. The progressivist synthesis was never dented. De la Beche ridiculed Lyell in a number of cartoons. The best known of these showed a class-room gathering of Liassic reptiles, re-emerged from the cyclical past of Lyellian earth history, listening to a professorial ichthyosaur (probably Lyell himself) lecturing on the anatomy of the human skull.[28]

Even Poulett Scrope, who was said to dislike Buckland and to sympathize with Lyell, took issue with the latter when it came to progressive change. In a review of Lyell's third volume for the *Quarterly Review* (1835) Scrope summarized the essential points of disagreement between the English school and the new Huttonians: 'Mr Lyell mistakes the essential character of his own argument. It is not the constancy of the laws of nature which he is contending for; this no one disputes. His real theory is, that there has been no progressive variation in the intensity of the forces which modify the earth's crust—but that a cyclical succession of such changes, of equal amount in equal periods, has been going on throughout all time, so far as geology enables us to explore its abysses.'[29] This, Scrope believed, ran counter to all analogy, which makes it a priori probable that the earth has had a beginning, will have an end, and that its history has been progressive.

## THE SMITH CULT

At about this time Smith became the centre of a cult which professed that he was 'the Father of English Geology'. In 1831 the first Wollaston Prize was awarded by the Geological Society under the presidency of Sedgwick. The council of the Society, dominated by members of the English school, granted the prize to Smith 'in consideration of his being a great original discoverer in English Geology; and especially for his having been the first, in this country, to discover and to teach the identification of strata, and to determine their succession by means of their embedded fossils'.[30] When Sedgwick presented the award to Smith he felt himself compelled

[28] See Rudwick, 'Caricature as a Source for the History of Science', *Isis* lxvi (1975), 534–60.

[29] liii (1835), 447–8.

[30] Extract from the council minutes, 11 Jan. 1831, *Proc. Geol. Soc.* i (1834), 271.

'to place our first honour on the brow of the Father of English Geology',[31] and Smith's reputation has continued to be based on this.

However, the Smith cult was not an attempt to set the historical record straight, but was part and parcel of the self-definition of the English school, nearly as artificial as Conybeare's choice of Leibniz as a forerunner of the progressivist synthesis. Smith possessed the perfect combination of attributes to serve as the mascot of the English school. His work on strata and fossils had been entirely based on observations made in the course of surveying and engineering work, unencumbered by prejudice and theory.

The occasion at which Smith was given the Wollaston Medal was the Oxford meeting of the British Association in 1832 under the presidency of Buckland. Murchison, as the new president of the Geological Society, presented the medal during a ceremony in the shrine of the academic establishment, the Sheldonian Theatre, and he sanctioned the epithet 'Father of English Geology'. At that meeting Conybeare specifically linked Smith with the English school. 'This school', he commented in his *Report*, 'generally recognizes the masterly observations of Smith, first made public in 1799, as those which have principally contributed to its establishment'.[32] This choice, by the geological establishment at Oxford and Cambridge, of a provincial ancestor enhanced its image as an empirical, fact-based school of geology, in contrast to the Huttonians, but it was hardly a reflection of historical reality. No such person as a single 'Father of English Geology' ever existed; geology was pre-eminently a collective endeavour, and if a single choice of progenitor of the subject had to be made from among its early contributors, men like Greenough and Parkinson would be most deserving of the title.

Smith's work, which had indeed been early and original, was nevertheless effectively bypassed during the formative years of the 1810s and 1820s. As late as 1826 Lyell, in his review of English geology, had not even mentioned Smith or his contributions. Fitton had given Smith some credit in the *Edinburgh Review* of 1818 and again in an enlarged republication of this early paper, serialized in the *Philosophical Magazine* (1832–3). But his assessment of Smith's actual influence was ambiguous; he seemed to imply that the

[31] Ibid. 279.
[32] *Report BAAS, 1831, 1832*, 370–1.

informal dissemination of Smith's results was of secondary importance as compared to the formal influence from abroad, especially Cuvier's. As for the latter's *Essai sur la géographie minéral-ogique des environs de Paris*, Fitton maintained 'that no publication has given a greater impulse to geological science;—bringing into view distinctly, and for the first time, that great class of deposits which connects the secondary strata with the products of still subsisting operations, establishing on impregnable ground the importance of zoological inquiries to the history of the earth, and affording some of the most masterly examples of the investigation of local details.'[33]

NOT A SCHOOL OF CATASTROPHISM

The belief that Buckland was the architect of a catastrophist synthe-sis and, moreover, that English geology represented a school of catastrophism is universally held among historians of nineteenth-century science.[34] Catastrophist geology constitutes the logical antithesis to the uniformitarian thesis of Hutton and Lyell, and Buckland and his school have been chosen for this antithetical role. Much of nineteenth-century intellectual history has been written within the framework of this polarity.

However, no such thing as a catastrophist synthesis ever existed in early nineteenth-century England. Neither Buckland himself nor English geology as a whole can be accurately or fairly called catastrophist. The English school was one of historical geology. Its cognitive identity was centred on the study of rocks and fossils as archives of earth history, and the synthesis it produced was that of progressivism, which emphasized the continuity of progress and the undisturbed length of geological periods rather than catastrophic interruptions.

The English school did believe in catastrophies, but more, in the early years, as a confirmation of written history and, in later years, as observational realities of discontinuity. Cataclysmal events seemed to be indicated by the extensive conglomerate deposits found at the terminations of major geological systems. The main interest of these deposits was not the evidence they provided for any mechanism of catastrophic devastation, but as punctuations of

---

[33] 'Notes on the History of English Geology', *London and Edinburgh Phil. Mag. and Journ. Sci.* ii (1833), 53.

[34] E.g. Gillispie, *Genesis and Geology*, 1959, ch. 4.

earth history which facilitated its periodization. Conybeare argued that four major conglomerate boundaries can be recognized, the oldest just above the Transition series (Old Red), the second above the Carboniferous system (New Red), the third above the Chalk (gravels of the Plastic Clay), and the last the diluvial gravels.[35]

There was, however, no preoccupation with any hypothesis as to the cause of such catastrophic events. Catastrophism never was to Buckland what uniformitarianism was to Lyell. To the latter uniformity was the central theme of his *Principles*, but the main argument of Buckland's *Reliquiae* and even more of his Bridgewater Treatise was that the majority of sedimentary rocks did not have a cataclysmal origin but had accumulated gradually during long periods of geological quiet. Buckland's interest was directed away from the catastrophism of Mosaical geology, and his fiercest opponents were the biblical literalists who recognized that his intention was to play down the geological consequences of Noah's deluge. To characterize the English school as catastrophist is a distortion based on a partisan overemphasis of a subsidiary element in its theory, partisan to Huttonian theory, by contrast to which English catastrophism was invented. It fails to give a true picture of its identity.

Nothing illustrates more unambiguously Buckland's leaning away from catastrophism towards a belief in long periods of geological quiet than his writings on the subject of fossil trees in an upright position and on the related question of the origin of coal beds. In a number of geological formations, but most frequently in the coal measures, fossil tree stems (dendrolites) occur, not only in a fallen horizontal position, but also in vertical or inclined attitudes relative to the bedding planes of the surrounding strata. Some of these trees died and became fossilized on the very spot where they grew (*in situ*), others were moved from their place of growth and deposited elsewhere, their heavy and bulky bases subsiding first. Many of the dendrolites are well preserved, indicating that the surrounding sediment layers were deposited before the stems began to rot. The size of some fossil tree stems is quite considerable: a height of several metres is not uncommon and examples are known of stems which measure fifteen metres or more in length.

An integrated bibliography on the subject of upright dendrolites goes back as far as the late 1810s. Enough examples were known by

[35] *Phil. Mag. Ann. Chem.* ix (1831), 191.

1819 to allow Jakob Nöggenrath to publish a monograph *Ueber aufrecht im Gebirgsgestein eingeschlossene fossile Baumstämme*; less than thirty years later the paleobotanist H. R. Göppert wrote an *Abhandlung* (1848) on the question of the origin of coal in which he listed over 250 examples of upright dendrolites from Europe and America.[36] No general geology textbook was without one or more pictures of these occurrences; de la Beche's *Geological Observer* and Lyell's *Elements of Geology* are good examples. A number of very large dendrolites fired the public imagination, and such 'dinosaurs of the vegetable kingdom' were prominently displayed in museums of natural history.

An international controversy as to whether or not the upright dendrolites were fossilized where they had grown began in 1818 with an exchange of letters in the *Bibliothèque universelle des sciences*. George MacKenzie from Scotland reported an example of a tree-stump *in loco natali*; Jean de Charpentier of Switzerland countered with an instance of a transported tree stem.[37] The issue came to the attention of the geological establishment when Brongniart wrote a paper 'sur des végétaux fossiles traversant les couches du terrain houiller' (1821); he described a subsequently classic case of tree stems in a coal and iron ore mine at Treuil near St. Étienne which looked like a fossilized forest: 'C'est une véritable forêt fossile de végétaux monocotylédons, *d'apparence* de bambous ou de grands *equisetum* comme pétrifiés en place.'[38]

Prévost, however, who did not agree with Cuvierian alternations of dry land with sea, objected to this interpretation, maintaining that the vertical position of the stems did not prove that they were *in loco natali*, and that the tree stems in the mine at Treuil had been transported; the lower ends of the stems were not at one and the same level and, in addition, no evidence of a forest soil was present.[39] An example where trunks were rooted in what appeared to be a fossil soil was the Portland 'Dirt Bed'. Buckland's study of it remained for some time the most convincing instance of fossil trees *in situ*. The study was frequently referred to during the 1830s, and

[36] It represented a prize-winning essay about the origin of coal written for the Hollandsche Maatschappij in Haarlem.

[37] viii (1818), 256–8; ix (1818), 254–8.

[38] 'It is a real fossil forest of monocotyledonous plants *looking like* bamboos or large *equisetum* converted into stone where they grew.' *Annales des mines* vi (1821), 362.

[39] 'Les continens actuels ont-ils été, a plusieurs reprises, submergés par la mer?', *Mém. Soc. Hist. Nat. Paris* iv (1828), 249–346.

as late as 1839 Buckland defended his interpretation against Prévost who had continued to object to *in situ* phenomena and who raised the issue at a special meeting of the Société gélogique de France at Boulogne-sur-Mer to which Buckland and other members of the Geological Society had been invited.[40] Agassiz also took exception to Buckland's emphasis on evidence for trees *in situ*. In his German translation of the Bridgewater Treatise Agassiz commented that he knew of only transported dendrolites and that those in the Portland 'Dirt Bed' were too close together to be a forest.[41]

The conclusion of rapid sediment deposition, based on the state of preservation of the fossil stems, was drawn by Bakewell in his textbook on *Geology* (1828). He stated 'that the strata were deposited rapidly, before the decomposition of the stem could be effected'.[42] There were other arguments for rapid sedimentation, derived from the occurrence of a number of dendrolites in the building-stone quarry of Craigleith near Edinburgh. One specimen in particular, discovered in 1830 and described by Witham, became famous. The stem, which had neither roots nor branches, was some fifty-nine feet long and slanted through a series of successive strata (figure 12). In his *Internal Structure of Fossil Vegetables* (1833) Witham argued that this and similar examples were evidently not *in situ* but represented instances of transport.[43]

Fairholme wrote immediately to the *Philosophical Magazine* to argue that he saw 'an insuperable argument against the theory of a *slow* deposition, in undefined periods of great extent, in these entire trees which intersect various parts of the coal-measures'.[44] In his *Mosaic Deluge* (1837) Fairholme referred to the Craigleith example when he wrote:

a tall stem, without either roots or branches, deposited in a vast mass of sedimentary matter, at an angle of only 45° with the lie of these strata, is a stronger testimony to the *rapidity* in the deposition, than even an *upright* stem; for while the latter might be supposed to have been capable of retaining an upright position, in a semi-fluid mass, for a long time, by the mere laws of gravity, the other must, by the very same laws, have fallen,

[40] *Bull. Soc. géol. France* x (1839), 427–31.
[41] *Geologie und Mineralogie,* ii, plate 57, 2.
[42] 168.
[43] See Christison, 'Notice of Fossil Trees', *Trans. Roy. Soc. Edinburgh* xxvii (1876), 203–21.
[44] 'Some Observations on the Nature of Coal', *London and Edinburgh Phil. Mag. and Journ. Sci.* iii (1833), 249.

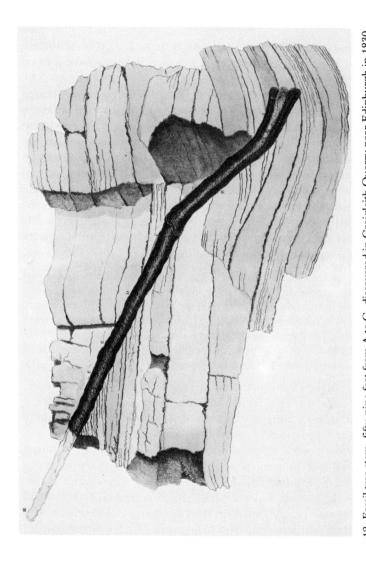

12. Fossil tree stem, fifty-nine feet from A to G, discovered in Craigleith Quarry near Edinburgh in 1830.

from its inclined, to a horizontal position, had it not been retained in its inclined position, by the rapid accumulation of its present stony matrix.[45]

Fairholme concluded that, as some beds had been accumulated rapidly, the entire geological column was likely to have formed during a giant cataclysmal event.

If Buckland had been a catastrophist he would have been pleased with the dendrolitic evidence for transport and rapid sediment deposition. But it worried him, for obvious reasons. The progressivist synthesis was based on a long stretch of geological time and on climatic change, the evidence for which was significantly based on *in situ* tropical fossils in northern latitudes. If such fossils had been transported, possibly over long distances, they would not constitute valid evidence of climatic change. Buckland admitted, reluctantly, that drifted plant fossils do occur; but the focus of his interest was decidedly on *in situ* instances. He admitted that some layers of sediment may have been deposited rapidly.

But, supposing it proved that one, or many thick sandstone beds of the coal formation were accumulated rapidly, this would not destroy the evidence of time we find in other strata. What is the proportion of one, or of twenty strata, when compared with the total known thickness of sedimentary formations. The greater part of these remain still loaded with the exuviae of animals which cannot be explained without admitting the lapse of long periods of time during their gradual deposition.[46]

Around 1840 the construction of railways through the Pennine coal field, north and south of Manchester, brought to light a number of new examples of upright dendrolites. Some particularly interesting instances, reported in cuttings near Bolton and Chesterfield, were described to the Geological Society during Buckland's second term as president. In his anniversary addresses Buckland gave a fair amount of space to the Carboniferous discoveries. In his first address (1840), he displayed an even-handed judgement and commented on the hypothesis of one author who believed all upright dendrolites to be *in situ*: 'In denying altogether the presence of drifted plants, the opinion of the author seems erroneous; universal negative propositions are in all cases dangerous, and more especially so in geology.'[47]

[45] 393–4.
[46] File 'Coniferae' OUM, Bu P.
[47] 21 Feb. 1840, *Proc. Geol. Soc.* iii (1842), 240.

In his second address (1841) Buckland returned to the subject of recent coal-measure discoveries which concerned not only fossil trees, but also so-called under-clays. Observations by de la Beche and by William Logan revealed that below many coal beds there is a fine-grained layer, characterized by the abundance of *Stigmaria ficoides*. Lindley and Hutton had interpreted this fossil as a separate plant, and Buckland now speculated that coal beds might have originated, not only from transported driftwood as he previously believed, but *in situ* by the colonization of mud flats or estuaries by *Stigmaria*, followed by such main coal plants as *Sigillaria*. His view soon received support from the discovery that *Stigmaria* is not a separate plant but the root system of *Sigillaria* trees.

This theory brought the origin of coal beds and of upright dendrolites together in a continuous sequence of *in situ* events. Buckland commented: 'We may explain the frequent occurrence of erect trees immediately above the upper surface of a bed of coal, as the cases we have spoken of near Bolton and Chesterfield, by supposing the roots of these trees to have found support and nutriment in the entangled remains of other plants which had preceded them on the same spot.'[48]

Whewell, in his *History of the Inductive Sciences*, introduced the distinction between 'the two antagonist doctrines of geology', one of 'geological catastrophies', the other 'geological uniformity'.[49] Being himself a member of the English school, he did not make the error of choosing Buckland as a contemporary representative of catastrophism. He cited Élie de Beaumont, and in his anniversary address of 1839, in which he reiterated the distinction, he also mentioned Murchison's belief in 'paroxysmal turbulence' during the early part of earth history.[50] Whewell's historiography was an early example of the history of ideas. It had a tendency to look at intellectual accomplishments as disembodied ideas, to be studied, classified, and contrasted on the bases of their inherent rationality. Thus catastrophism served as a contrast to Lyell and the Huttonians, whose theory can be fairly characterized as uniformitarian.

The history of ideas has given the wrong answer ('catastrophism') to the question: 'What was the cognitive identity of the English

[48] 19 Feb. 1841, ibid., 492.
[49] iii (1837), 606–24.
[50] 15 Feb. 1839, *Proc. Geol. Soc.* iii (1842), 92.

school?' The right answer is 'historical geology'. Moreover, the history of ideas can provide no answer at all to the question which should follow: 'Why was the geology of Buckland and his school pre-eminently historical?' Here the answer does not lie in the inherent rationality of the subject of historical geology, but in the social milieu of the major members of the English school.

### HISTORIANS OF THE SUBTERRANEAN WORLD

The central figures of the English school were clerical academics. Buckland, Conybeare, Sedgwick, and Whewell were educated in the classical system and took holy orders. At the old English universities they lectured to students many of whom were destined for the Church. In order to get the new subject of geology accredited they had to align it with the existing educational tradition. The historical aspect of geology was eminently well suited for this, in contrast to the economic aspect, for example, which was thought not to merit academic rank, and to be best left to provincial and metropolitan institutions. By virtue of their education and occupation the central members of the English school had an aptitude for the study of earth history as an extension of and complement to traditional world history.[51]

Thus the academic setting of English geology enhanced its character as a historical subject. It forced upon Buckland and his circle an obligation to relate geology and its new discoveries to such facts of sacred and world history as creation and deluge. Buckland's early diluvialism had been a response to this contextual pressure. By about 1830 emphasis began to shift from the deluge to the creation story. Buckland was forced to attempt a reconciliation of the new perspective of progressive earth history with the biblical account of creation. This was at the same time an attempt to give geology more independence at Oxford and Cambridge. Sedgwick's *Discourse*, which presented a concise summary of the progressivist synthesis, was also a plea for educational reform. Buckland's Bridgewater Treatise was a comprehensive compendium of historical geology, but it contained also his major reconciliation scheme.

This contrasted with the place of geology in Scottish and many continental universities. To Brewster, Fleming, and other Scottish

[51] See chs. 1 and 4.

Evangelicals the subjects of geology, history, and theology were autonomous realms of study; analogies between these might be enlightening, but their substances should not be mixed. This was also Lyell's position. These men were not irreligious; on the contrary. But they worked under no academic obligation to bring their results into accord with traditional scholarship and belief. Agassiz commented from his continental vantage-point that the clerical patronage of geology in England was the reason why the English school showed a preoccupation with the relationship of geology with Genesis.[52]

The interaction of geology with history and theology at Oxford during the 1830s was apparent on various levels. Buckland recurrently justified his subject as a subterranean extension of human history: 'If it be interesting to trace the history of our country during the Norman, or Saxon, or Roman, or Celtic periods of its history, it is not less interesting to extend our investigations still further backwards into yet more distant periods of entirely different operations at the bottom of the sea when the strata of our present land were themselves receiving their formation.'[53] Historical metaphors were frequently used in geology. Murchison, the most eminent convert to geology from outside the academic establishment, recruited by Buckland and Sedgwick, wrote: 'Rocks are to the Geologists what *Papyri* are to the Antiquary, imparting to every one who diligently lays them open, the history of the ages that have preceded us.'[54]

The preoccupation of these men with the dovetailing of geology and history rubbed off on their families. A letter by Charlotte Murchison to Mary Buckland in 1832 illustrates this: Murchison was president of the Geological Society and Oxford had just appointed Horace Wilson as its first professor of Sanskrit:

The President requests you will consult the 114th psalm for a fine poetical version (tho' *astonishingly correct*) of what has happened in one of the revolutions of the surface of the globe. And he begs the *D.D.* will rub up the books of his Hebrew and study geology in Job and Solomon. He also requests you will get your new eastern professor to enlighten you with translations of the geological phenomena recorded by Mohammed Kazwini, the

[52] *Geologie und Mineralogie*, i, 37.
[53] Bridgewater Treatise notes, OUM, Bu P.
[54] *Outline of the Geology of the Neighbourhood of Cheltenham*, 1834, 3.

BounDehesch of Zoroaster etc. (Charlotte added:) 'Lyell can never stand against all this cosmogony!'[55]

However, an increasing competitiveness grew, during the 1830s, between the new geology and traditional history. Nares, the regius professor of modern history, wrote *Man, theologically and geologically* (1834). A wave of traditionalism swept Oxford when the Tractarian Movement was set in motion, when the Hampden affair aggravated existing divisions, and when student attendance at the science lectures began to drop alarmingly. Buckland and his circle reacted by asserting the intellectual integrity of their research and by an attempt to influence senior appointments, especially to the regius chair of history. One lecture fragment by Buckland reads:

Since the beginning of the present century, geological investigations have been conducted with so much accuracy and caution that many of the results which have been arrived at, may be regarded as legitimate conclusions from facts established on the concurrent evidence of independent observers in various countries; these facts, when arranged carefully according to the chronological order of succession in which they are presented by nature, have become the foundation of a branch of history of the highest and most venerable antiquity recording transactions that took place upon the surface of our planet antecedent to the period at which all human records begin, a history which though unauthenticated by a single manuscript, or by the tittle of human record, is nevertheless attested by documents not less intelligible and much more faithful then those on which the historiographer of the human race rests the evidence of events which constitute the materials of his science. Geological evidences are, in fact, more nearly allied to those which the researches of the antiquary brings in aid of the historiographer and which often constitute the best foundation of his knowledge. The discovery of ancient coins and medals and of inscriptions engraved on tablets of marble and brass are justly considered to afford the most convincing evidence of events which took place in the distant centuries to which they relate; still more perfect is the evidence of those most faithful of all documents, the medals of nature, which record in characters not less intelligible the successive histories of events that occurred on the surface of the earth before the creation of the human race.[56]

Buckland fully approved of Nares's successor in 1841, Thomas Arnold, headmaster of Rugby. Arnold had been one of Buckland's early students, and had objected to Hampden's persecution.

[55] 9 Oct. 1832, OUM, Bu P.
[56] Miscellaneous notes, OUM, Bu P.

Arnold's biographer quoted: 'When Professor Buckland, then one of our Fellows, began his career in that science, to the advancement of which he has contributed so much, Arnold became one of his most earnest and intelligent pupils, and you know how familiarly and practically he applied geological facts in all his later years.'[57] Arnold died early and his successor was to be chosen by Peel. Buckland and Conybeare, in concert with Whewell, plotted the choice of a man who would be sympathetic to their cause, namely James Prichard, the physician and ethnologist whose *Researches into the Physical History of Mankind* (1813; 1826) was closely allied to historical geology in its approach to human history. Buckland actually suggested that the regius chair of modern history be used to teach ethnography.[58] Prichard, who had a Bristol Quaker background, was not, however, appointed, and the chair went to the safe but unremarkable classicist John Cramer.

Buckland's empathy with Prichard was partly based on the latter's view of the Bible as pertinent to human affairs only, not to natural history. This narrowed conception of the relevance of Genesis was intended to cut the umbilical cord between the Bible and the study of nature, as this had existed in the *origines sacrae* tradition. Prichard had asked rhetorically: 'And of what importance could it be for men to be informed (in the Bible) at what period New Holland began to contain kanguroos, and the woods of Paraguay ant-eaters and armadillos?'[59] This interpretation aided geologists in their quest for freedom from interference by the humanities and by theology in particular.

A similar restriction in scope of the validity of the Bible was advocated by Sedgwick in his anniversary address of 1830. Baden Powell emphatically agreed. Hampden too supported this down-grading of the Bible's significance when he suggestively set the following subject for the divinity school examination: 'The Scriptures have not been designed to convey philosophical instruction.'[60]

Few people were better placed than Buckland to address themselves to the delicate matter of the territorial disputes between geologists and biblical scholars. He was a leading scientist; he

---

[57] Stanley, *Life and Correspondence of Arnold*, 1844, 10.
[58] Buckland to Whewell, 22 July 1842, TC, Wh P.
[59] *Researches into the Physical History of Mankind*, i, 4th edn. 1841, 101.
[60] Miscellaneous notes, OUM, Bu P.

occupied an ecclesiastical position of some weight; he was patronized by conservative politicians and church dignitaries; and he had distinguished himself in this task in his earlier *Reliquiae Diluvianae*. His Bridgewater Treatise, begun in 1832 and delayed till 1836, was therefore eagerly awaited. In it Buckland completed the disentanglement of earth and human history, begun around 1830, by pushing back the last geological débâcle from the historical time of the Mosaic deluge to the period immediately preceding the creation of man. From the outset he emphasized that the language of rocks and fossils are as much a divine revelation of truth as the language of the Bible. In a lofty introduction Buckland stated that the facts of geology 'make up a history of a high and ancient order, unfolding records of the operations of the Almighty Author of the Universe, written by the finger of God himself, upon the foundations of the everlasting hills'.[61]

The main discrepancy between the language of geology and that of the Bible existed, he believed, in 'the disclosures made by Geology, respecting the lapse of very long periods of time, before the creation of man'.[62] To remove this apparent inconsistency Buckland emphasized that biblical coverage did not stretch beyond human history, and that a proper exegesis of the first two verses of Genesis would provide a means to connect earth history to human history, allow geology all the time it needed and leave a literal interpretation of Genesis entirely intact. He wrote:

The disappointment of those who look for a detailed account of geological phenomena in the Bible, rests on a gratuitous expectation of finding therein historical information, respecting all the operations of the Creator in times and places with which the human race has no concern; as reasonably might we object that the Mosaic history is imperfect, because it makes no specific mention of the satellites of Jupiter, or the rings of Saturn, as feel disappointment at not finding in it the history of geological phenomena, the details of which may be fit matter for an encyclopedia of science, but are foreign to the objects of a volume intended only to be a guide of religious belief and moral conduct.[63]

Thus the biblical chronology had never been intended to fix the age of the earth, only that of the human species. There was no

[61] *Geology and Mineralogy*, i. 7–8. See Sedgwick's anniversary address of 19 Feb. 1830, *Proc. Geol. Soc.* i (1834), 207.

[62] Ibid., 8.

[63] Ibid., 14–15.

reason to expect to find earth history recorded in the Mosaic account of creation. Buckland rejected the notion that the days of the creation week in Genesis were periods of geological time. This interpretation had been suggested by Parkinson (1811, 1822), Townsend (1813), and Kidd (1815); Mantell had promoted it (1822); many others followed it, and Buckland himself had considered it in his *Vindiciae*. It had, however, two disadvantages. First, it limited the freedom of geology in that the sequence of geological periods would have to be identical with that of the creation days (which it is not); and second, a literal meaning of the word 'day' would have to give way to a symbolic interpretation.

Buckland promoted another exegesis, also previously considered in his *Vindiciae*, namely that the first verse of Genesis, 'In the beginning God created the heaven and the earth', is not a prospective summary of the creation week in the following verses, but a retrospective reference to the primeval creation of matter, the stars, the planetary system, and also the earth. The first part of the second verse, 'And the earth was without form and void', takes up the history of the earth after an indefinite and possibly very long interval at the moment of the destruction of the last geological world, as a preparatory statement to the creation of the human world. Buckland added his famous conclusion: 'millions of millions of years may have occupied the indefinite interval, between the beginning in which God created the heaven and earth, and the evening or commencement of the first day of the Mosaic narrative.'[64]

This interpretation was supported by a long footnote by Pusey, regius professor of Hebrew at Oxford, in which he discussed the various words used in the original for 'to create'. This allowed Buckland to suggest that the so-called creation of sun, moon, and stars on the fourth day of the creation week was not a *creatio ex nihilo* but a rearrangement, probably of atmospheric conditions, which made the heavenly bodies visible and useful to man. The contribution from Pusey was a good political move, especially in the Oxford context. Buckland wrote to Sedgwick: 'I have not much fear for my theology, having shewn my early sheets to the Bishops of Chester and Llandaff, and to Professors Burton and Pusey, all of whom are perfectly content.'[65] Not only allies like Copleston and Sumner, but also the Evangelical Faber supported Buckland's reconciliation scheme.[66] It was very successful; the exegesis gave maximum freedom to geology and left the literal meaning of the creation days intact. However, it met with disapproval from Tractarians like Keble and other traditionalists. But Buckland attempted to ignore their obscurantism:

It was a matter of course that I should incur the censure of those who doubt the reality of the facts disclosed by Geology. I am well content to endure their jibes, and the reproof or incredulity of the over-zealous in religion, who take refuge in believing that the Deity if He pleased, might have created in an instant, all things as they are; as if they would say, God if He pleased might have made mummies as they are, and that there is no reason to believe these mummies to have ever been living men, that advanced to maturity, through stages analogous to those which mark our own progress from infancy to age. To this class of persons I do no harm; they adhere to their belief in Scripture and suffer nothing from their incredulity in geology; to all reasonable persons, I rejoice to find I have given entire satisfaction and I hope I have done much good by shewing to the multitudes who saw a difficulty in reconciling Geology with Theology, that there is no discrepancy at all between them. It was high time this should be done, and not only from England but also from France have I here received many letters expressing satisfaction at the manner in which that part of my subject has been treated.[67]

An identical or similar exegesis to Buckland's had been expressed by Chalmers in his *Evidence of the Christian Revelation* (1814) and by a number of other divines; Higgins had suggested it in his *Mosaical and Mineral Geologies* (1832), Sedgwick had alluded to it, and the *Christian Observer* had discussed it. Buckland's presentation, however, gave the interpretation a new gloss of quality and respectability. He was supported and followed by a large variety of people; men like Pye Smith and Gibson popularized it; even Babbage, rather peremptorily, approved of it as a partial victory of science over theology. The *Quarterly Review* intoned authoritatively about the subject of geology:

If there are any persons yet deterred from the study of this fascinating science by the once prevalent notion, that the facts, or theories if you will, that it teaches, tend to weaken the belief in revealed religion, by their

[66] Supplementary Notes, *Geology and Mineralogy*, 2nd edn. 1837, i. 597. See also 'Geology and Holy Scriptures', *Mag. Popular Sci.* ii (1830), 465–8.
[67] Buckland to Irvine, 25 Feb. 1837, CC, Bu C.

apparent inconsistency with the scriptural account of the creation and early history of the globe, —*here*, in a work of a dignitary of the church, writing, *ex cathedra*, from the headquarters of orthodoxy, they will find the amplest assurances that their impression is not merely erroneous but the very reverse of the truth.[68]

Some authors construed Buckland's exegesis as a declaration of geology's independence, a sedition against the sovereignty of its ecclesiastical patronage. For example, Brewster wrote about the notion of prehistoric geological periods: 'This is now the universally received doctrine of the English School; and such has been the progress of liberal opinions that, in assemblies composed of Churchmen, and Dissenters, and Conservative statesmen, we have heard the walls ring with rapturous joy, when geology renounced her ecclesiastical tenure, and demanded a lease of MILLIONS OF MILLIONS of years for the range of their enquiries.'[69]

However, this presumed renunciation was Scottish wishful thinking; the content of historical geology was in part structured by the very preoccupation with the relationship of earth history with Mosaic history, and ecclesiastical patronage of geology was to continue for some time. Buckland, for one, moved even closer to the Church when he became dean of Westminster. Brewster, and many others, understandably resented the privilege and exclusiveness of the Anglican patronage of geology at the ancient English universities. This may go some way to explain the popularity of Lyell, as a rallying-point of geologists outside the circle of clerical academics.

Lyell's theoretical position was not shared by a majority of his colleagues in England.[70] But he represented geology without either the privilege or the constraints of its Anglican tenure at Oxford and Cambridge. While Buckland and Sedgwick were under an academic obligation to reconcile geology with biblical history, Lyell could ignore the latter as of no consequence to his subject. Lyell's *Principles* included no reconciliation scheme, and about the Mosaic deluge he remarked: 'For our own part, we have always considered the flood, if we are required to admit its universality in the strictest sense of the term, as a preternatural event far beyond the reach of

[68] lvi (1836), 31. The *Literary Gazette* was also complimentary, 12 Nov. 1836, 721–2.
[69] *Edinburgh Review* lxv (1837), 13.
[70] Bartholomew, 'Non-progress of Non-Progression', *Brit. Journ. Hist. Sci.* ix (1976), 166–74.

philosophical inquiry, whether as to the secondary causes employed to produce it, or the effects most likely to result from it.'[71] This contrast between the clerical and the secular control of science was central also to the famous debate about evolution at Oxford when Huxley clashed with Wilberforce, one of Buckland's early pupils.[72]

[71] iii, 275. See also Rudwick, 'The Strategy of Lyell's *Principles*', *Isis* lxi (1970), 4–33.
[72] See Lucas, 'Wilberforce and Huxley', *Hist. Journ.* xxii (1979), 313–30.

# 16
# Traditionalist Opposition

## CHRISTIAN ESCHATOLOGY

The opposition to geology by biblical literalists within Oxford did not form an unbroken continuum with that in England at large. Inside the university the Oxford Movement was philo-Catholic, but outside opposition tended to be of a self-confessedly anti-Papist character. However, neither inside nor outside Oxford was the conflict a straightforward question of geology versus Genesis. Even though contemporary journalism did reduce it to this oversimplification, it is intellectual indolence for later historians to have done the same.[1] After all, the leading geologists were as much clerical and devoutly religious as they were scientific. Buckland's reconciliation scheme gave full credibility to both geology and the text of Genesis. However, to each a different stretch of history was allotted. This separation of earth history from biblical and human history encountered fierce opposition from many biblical literalists, because it invalidated their eschatological world-view, together with its scholarly basis and its contemporary political and social consequences.

In the traditional perspective of world history the physical existence of the earth was limited to the time-span of human history. The past of the earth and that of man were not seen as separate realms, but were believed to be intimately interwoven to produce a prospect of eschatological anticipation. The physical vicissitudes of the heavens and the earth, the creation of the world and its destruction by water in the deluge were thought of as constituting an expression and portent of human destiny. On the reality of these events depended the veracity of a belief in the Second Advent of Christ, the imminence of his millennial reign, the destruction of the earth by fire, and the creation of a new heaven and a new earth. Small-scale phenomena such as earthquakes,

[1] E.g. Chadwick, *Victorian Church*, i (1971), 559.

comets, and other signs in the sky were regarded as omens of the ultimate conflagration.

This tradition of world history had deep roots in English culture. Apocalyptic theology and especially millenarian expectations had been popular as late as the seventeenth and eighteenth centuries.[2] But in the course of the first half of the nineteenth century a considerable amount of new enthusiasm was generated for the apocalyptic belief. The various beastly symbols in such Old Testament prophecy as Daniel or the New Testament Revelation of St. John found new social and political meaning in events which ranged from the revolution in France to the European upheavals of 1848. The rise to imperial power of Napoleon and his likeness to the apocalyptic beast almost eclipsed the evil image of the Pope as the Antichrist, though the Catholic Emancipation Act of 1829 and the 'Popery' of the Oxford Movement in subsequent decades re-established the danger of the Church of Rome as the apocalyptic harlot.

This conservative Protestant world-view rested on the pillars of sacred history and the study of prophecy. These subjects had been part and parcel of the classical system at England's old universities, and they remained academically viable through the first half of the nineteenth century, as shown by a variety of publications. Examples are the several Oxford University Press editions through which Stillingfleet's *Origines Sacrae* went; the *Fasti Hellenici* by Henry Clinton, the first volume of which contained a large appendix on sacred chronology; and the *Connection of Sacred and Profane History* by Michael Russell, who intended his book to be a sequel to those of Prideaux and Shuckford, two great names of eighteenth-century *origines sacrae* scholarship. Although the authors of these and similar academic works were not necessarily themselves among the vociferous prophets of an imminent apocalypse, their contributions to sacred chronology sustained the basis on which rested the predictions of the end of the world and its precise date.

Prophecies of this kind were earnestly preached by many worthy churchmen, both Anglican and nonconformist. In London in the 1820s the Scotch Presbyterian minister Edward Irving attracted large crowds to his sermons, in which he expressed his belief in an imminent millennium. A ready-made platform for the voice of

---

[2] See Cohn, *Pursuit of the Millennium*, 1957; Hill, *Antichrist in Seventeenth-Century England*, 1971; Reeves, *Influence of Prophecy in the Later Middle Ages*, 1969; Toon, *Puritans, the Millenium and the Future of Israel*, 1970; Webster, *Great Instauration*, 1975.

prophecy—and doom—existed in the Warburtonian Lecture at Lincoln's Inn which had been established to prove the truth of the Christian religion from the completion of prophecy in the Old and New Testaments. A number of early nineteenth-century Warburtonian lecturers, such as Nolan and Elliott, added their voice to the apocalyptic chorus.[3] Elliott's later *Horae Apocalypticae* caused a considerable stir; his voluminous production contained an 'Apocalyptic Chart' which indicated that the millennium was near and that its inception was due before the end of the century, possibly as early as 1875.

Elliott was a member of the so-called Clapham Sect of conservative Evangelicals. Within the Church of England Evangelicals paid more than average attention to prophetic dates. When W. J. Conybeare wrote his famous article on church parties for the *Edinburgh Review* he commented sardonically that in Evangelical circles 'Novels and fairy-tales, it is true, are forbidden luxuries; but their place is abundantly supplied by the romantic fictions daily issuing from the Prophetic Press.'[4]

One of the most prolific apocalyptic writers of the period was William Cuninghame, initially allied to the Evangelicals and an early contributor to the *Christian Observer*. The dates of his numerous publications cover the entire first half of the nineteenth century. Cuninghame combined just about all the characteristic features of the apocalyptic imagination, namely an interest in sacred chronology with a belief that from it the beginning and true age of the earth could be computed, a strong opinion about the relative merit of the different textual versions of the Old Testament (Hebrew versus Septuagint), an interest in biblical prophecy with a conviction that from it the end and true duration of the world can be calculated, an explicit anti-Papism, a desire to convert the Jews and resettle them in Palestine, and a censorious and often vacuous rhetoric. The very long titles of his books were something of a summary of his views. Cuninghame's writings were filled with millenarian anticipation; initially he dated the Second Advent of Christ for 1841, but under force of circumstances he put forward the date to 1867. Writing in 1839 Cuninghame warned 'that the great day of the Lord is at hand', and he concluded:

[3] See Hunt, *Religious Thought in England in the Nineteenth Century*, 1896, 67 and *passim*.
[4] xcviii (1853), 296.

Lastly, when we advert to the stupendous events which are to fill the short period of 28 years, between the point of time where we now stand, and the termination of Daniel's period above-mentioned, which comprehend the Second Advent in glory of the Lord, the resurrection of his sleeping saints, the change and rapture of his living saints; the restoration of Israel in the flesh, the rebuilding of Jerusalem; the destruction of Babylon, and the judgment of God against all nations; the discomfiture of the confederacy of Gog, the treading of the wine-press of wrath in the day of Armageddon, and the burning of the body of the Beast, and establishment of the kingdom of Christ in glory: when we consider that this stupendous chain of events is to fill the short space of 28 years, we are persuaded that in whatever order these events are to develop themselves, which is at present in a great measure hid from us, the commencement of the mighty catastrophe, in whatever manner, and by whatsoever event it may begin, is at the very door; and believing that the scientific chronology of the year 1839 is given to us, and spread out before our eyes, for a sign of this, I send forth these pages commending them to the blessing of Him who has warned us of the sin and danger of not improving his talent.[5]

When these 'stupendous events' did not occur Cuninghame decided to help bring them about; in a *Letter* to Ashley, president of the Society for Promoting Christianity amongst the Jews, he urged the taking of immediate measures for the Jewish colonization of Palestine.

Cuninghame may have been somewhat of a maverick, but many took him seriously and either agreed with his apocalyptic dates or suggested alternative years for the end of the present order of things. Robert Wallace, a mathematics tutor at the University of London, followed Cuninghame in a *Dissertation on the True Age of the World* (1844).

This apocalyptic world-view shared no common ground with modern geology, in relation to either past or future. The past was, for both, the key to the future, and during the 1820s it had still been possible for the apocalyptic imagination to be fired by diluvial geology, even though a number of biblical literalists had recognized the incongruity of diluvialism and Old Testament history. But in the course of the 1830s the apocalyptic tradition and modern geology definitively diverged.

Officially the English school did not speculate about such non-empirical matters as future earth history. In his anniversary address of 1830 Sedgwick stated that the precise length of earth history cannot be defined 'and still less do we dare to speculate about the physical revolutions of the ages which are to come'.[6] However, the new perspective of the past did change anticipations of future events. If earth history had been characterized by long periods of progressive succession, then it was to be expected that the future would evolve along the same lines. Pidgeon wrote in 1830: 'Revolution has succeeded revolution—races have been successively annihilated to give place to others. Other revolutions may yet succeed, and man, the self-styled lord of the creation, be swept from the surface of the earth, to give place to beings as much superior to him as he is to the most elevated of the brutes.'[7] About a decade later the *Literary Gazette* asked if man had not progressed since the moment of his creation, and answered: 'Surely he has; and even yet may be but the link upwards to a higher gradation in the scale of being. Winged angels may inhabit our sphere before it utterly pass away; and all the knowledge of our day, anxious as we are to acquire it, groping between darkness and light, be to them the foolishness of children, and the whisperings of ignorance.'[8]

Thus the foundation of the Christian eschatological hope seemed to dissolve in the mist of a vast and progressive diorama of earth history. To the traditionalists Buckland's 'millions of millions of years' became the shibboleth of the new ungodly geology. Their favourite quotation was an ironical passage from William Cowper's *The Task* (1785):

> Some drill and bore
> The solid earth, and from the strata there
> Extract a register, by which we learn
> That he who made it, and reveal'd its date
> To Moses, was mistaken in its age.[9]

Members of the English school were well aware of the fact that the new perspective of earth history detracted from the uniqueness of man in nature and of his relationship to God, although they were

[6] 19 Feb. 1830, *Proc. Geol. Soc.* i (1834), 211.
[7] *Animal Kingdom*, 39.
[8] 14 Aug. 1841, 513.
[9] Bk. iii, line 150–4.

slow to discuss this. But Whewell, in his last major work on the *Plurality of Worlds* (anonymous, 1853), attempted to rescue something of the traditional belief that man, after all, might be the final product of creation and unique in both space and time.

Astronomy had appeared to diminish man's place in space, and geology now did the same in time. Whewell wrote: 'If the earth, as the habitation of man, is a speck in the midst of an infinity of space, the earth, as the habitation of man, is also a speck at the end of an infinity of time. If we are as nothing in the surrounding universe, we are as nothing in the elapsed eternity; or rather, in the elapsed organic antiquity, during which the earth has existed and been the abode of life.'[10] The geological discovery of a plurality of worlds in time gave new credibility to the old speculation of a plurality of worlds in space. William Jacob, an astronomer to the East India Company, argued that if habitable bodies exist across the visible universe, geology renders it probable that these exist in different progressive stages; that some of them are as yet devoid of life, that others are becoming clothed with vegetation, and inhabited by lower animals; and probably also that some of them have attained a level of development similar to that of the earth or even higher.[11]

But Whewell turned the argument for a plurality of worlds around. Because man has occupied only the last, very limited stretch of geological time, why should his place in space not be equally limited? Speaking about God, Whewell wrote: 'He has found it worthy of Him to bestow upon man His special care, though he occupies so small a portion of time; and why not, then, although he occupies so small a portion of space?'[12] In the last chapter of his book Whewell came to the crux: man is the head of creation, in his present condition; but is that condition the final result and ultimate goal of the progress of creation in the plan of the Creator? The trusted method of Baconian induction was of no use in trying to answer this question, and Whewell had to resort to the following self-glorification: 'When pure Intellect is evolved in man, he approaches to the nature of the Supreme Mind: how can a creature rise higher? When mere impulse, appetite, and passion are placed under the control and direction of duty and virtue, man is put under a Divine Government: what greater lot can any created

[10] 100.
[11] *A few more Words on the Plurality of Worlds*, 1855, 42.
[12] *Plurality of Worlds*, 103.

being have?'[13] One can understand those of Whewell's contemporaries who saw in his argument a disguised proclamation that the mastership of Trinity was the pinnacle of earthly existence. The logic of his argument was worthy of a clever undergraduate essay. It did not convince someone like Brewster, who wrote a refutation under the title *More Worlds than One* (1854).[14]

## MOSAICAL COSMOGONY

The wave of fundamentalist criticism which had followed the publication of Buckland's hyena den theory redoubled when, in the course of the 1830s, the progressivist synthesis emerged. Nolan, in his notorious Bampton Lecture at Oxford on the *Analogy of Revelation and Science* (1833), combined a belief in a universal deluge and in an imminent end to the world with an explicit disavowal of the theory of progressive succession in earth history. The Bible, Nolan asserted, speaks of only one creation, and he concluded: 'The notion of a progressive development of the earth by successive creations and destructions, must be therefore rejected, as not entitled to more serious respect, than any mythological fable which professes to account for its origin.'[15] Henry Cole, formerly of Clare Hall in Cambridge, agreed with Nolan. He addressed a long letter to Sedgwick, *Popular Geology Subversive of Divine Revelation* (1834), in which he reproached the politicians for having regarded Catholic emancipation as a question of political expediency, rather than of religion and revelation. Turning to Sedgwick, Cole continued:

And the very same device and delusion, Sir, is the same subtle destroyer now pursuing through the instrumentality of men of science, and especially by means of the science of GEOLOGY; and he pursues exactly the same course now, in the scientific, as then, in the political world,—deluding Geologists to conceive, and then to declare, that the science of Geology, and of the world's creation 'RESTS UPON ITS OWN BASIS!' and that 'it has nothing to do with divine Revelation, nor divine Revelation with it'.[16]

[13] Ibid., 270.
[14] See Brooke, 'Natural Theology and the Plurality of Worlds', *Ann. Sci.* xxxiv (1977), 221–86; Yeo, 'William Whewell', ibid., xxxvi (1979), 493–516. Whewell at one time had argued in favour of a plurality of worlds. See his Bridgewater Treatise on *Astronomy and General Physics*, 1833, 279–93.
[15] 109.
[16] 4.

A review of Cole's letter in the *Athenaeum* (1834) shows how vitriolic the polemics were becoming; the writer called Cole ignorant of geology and concluded that Cole's views 'constitute a case for his physician rather than his critic'.[17]

The wave of fundamentalist criticism rolled on. George Croly, a classical scholar like Nolan and Cole, continued the attack on the theory of progressive development as 'only a specimen of the rage of foreign philosophy for generalization. The fact is,' he continued, 'that those deposits are irregular in the extreme, are found under all varieties of circumstance, and the strata in which they lie are so few, and so widely interrupted, that it would be utterly absurd to consider them as forming any of the important integuments of the globe.'[18] The wave of criticism began to crest shortly after Buckland made his Bridgewater Treatise public, particularly in reaction to his views on the age and history of the earth, which he had presented at the Bristol meeting of the British Association (1836). Croly, writing in *Blackwood's Edinburgh Magazine* about the Liverpool meeting of 1837, commented savagely that 'at Liverpool we certainly were spared the offensive folly of the hurrah of the rabble of cognoscenti, on a clergyman's giddily giving a date and origin to the world wholly contradictory to that which is expressly given in the Bible.'[19]

The allusion to Buckland is unmistakable, and his reconciliation scheme came in for the lion's share of traditionalist criticism; newspaper articles, pamphlets, or even book-sized rebuttals of his views flooded the market. One pamphleteer saw Buckland's geology as 'a direct and real, though disavowed attack on the Mosaic narrative of the creation, made by a Clergyman of the Church of England', and in the English school he detected the influence of French science: 'So much of geology,—of its science and of its discoveries,—proceeds from the French school, that it is to be feared the *malaria* of French philosophy has sometimes mildewed the more healthy character of English science.'[20] Vacuous rhetoric, even worse than this, filled the pages of a number of other pamphlets.[21]

In 1838 the first of a series of pamphlets directed against the

[17] 11 Oct. 1834, 741.
[18] *Divine Providence*, 1834, 100.
[19] xlii (1837), 690.
[20] Brown, *Reflections on Geology*, 1838, 3, 38.
[21] E.g. Johnsone, *Vindication of the Book of Genesis*, 1838. Less hostile was Best, *After Thoughts on Reading Buckland's Bridgewater Treatise*, 1837.

English school came out, written by William Cockburn, a Cambridge man of some distinction and dean of York. It was a *Letter to Professor Buckland, Concerning the Origin of the World*, in which Cockburn strenuously objected to the notion of a long geological history prior to human history. Like Croly, he combined an attack on modern geology with one on the meetings of the British Association. In a *Remonstrance upon the Dangers of Peripatetic Philosophy* (1838) Cockburn denigrated its meetings: 'It appears to me that these annual assemblies of Thespian Orators, while they confer no benefit upon science, have been, and are likely to be, injurious to religion.' As an example he selected Buckland's work: 'If Mr. Buckland be right, Moses must be wrong.'[22] Cockburn spared none of the leading geologists; he addressed a subsequent pamphlet, on the *Creation of the World* (1840), to Murchison, in which he speculated that the Silurian rocks had been produced by submarine volcanism.

Cockburn's crusade against historical geology became involved in a major battle in 1844 when the British Association met in his home city of York. He used an opportunity to address the meeting to reiterate his criticism and suggest an alternative mechanism for the deposition of rocks and fossils. He believed himself so successful in this encounter that he boasted in a letter 'To the Inhabitants of York': 'The Philistines are beaten with the very weapons they had prepared against us, and the head of Goliath is cut off with his own sword.'[23]

Not all pamphlets against the English school were composed of merely censorious condemnation. A number of serious attempts were made to collect facts and work out a theory opposed to slow deposition and progressive succession over long periods of time. Facts in support of rapid sediment deposition were cited by Fairholme in his *Geology of Scripture* and in his *Mosaic Deluge*. Among these facts were the phenomena of upright dendrolites and the preservation of reptilian footprints on bedding planes. Fairholme speculated that the mechanism for rapid sedimentation had been strong tidal flows of the waters of the deluge: 'The waters of the whole sea must then have been, as we have before shewn, heavily

[22] 5, 7.
[23] *The Bible Defended against the British Association*, 1844, 23. See also Clark and Hughes, *Life and Letters of Sedgwick*, ii (1890), 76–80. Morrell and Thackray, *Gentlemen of Science*, 1981, 229–45.

charged with their preternatural burden; and every successive tide must, consequently, have deposited some additional beds upon the growing earth. In this manner alone, can we account for the rapid deposition of the trees we have just been considering; and, in this same manner alone, can we also account for the preservation of those animal foot-marks now discovered between the strata.'[24] William Rhind, also, used the examples of the Craigleith dendro-lites to argue against long periods of geological time.[25] Several others elaborated the tidal wave mechanism, especially in application to the coal measures.

The progressive occurrence of fossil types in the rock record was explained as a function of the spatial distribution of animals and their ability to flee the rising waters of the deluge, away from low areas to high ground. Nolan had already suggested this in his Bampton Lecture; but it was further worked out by Cockburn, most elaborately so in his *New System of Geology* (1849). He speculated that submarine volcanism during the antediluvian period had produced the Silurian rocks at the bottom of the oceans; that at the time of the deluge these volcanic eruptions had increased in intensity while rain poured down, and that their alternating effect had produced the systems above the Silurian. He concluded: 'The above hypothesis will explain, also, the order of succession in which we find the fossil remains of sea and land animals, in all the various formations. The creeping things at the bottom of the sea were the first destroyed; then the fish; next, the animals inhabiting the marshes near the sea; afterwards, the heavy quadrupeds that could not run from the rapidly increasing waters; and, lastly, the more active animals, which had for a time escaped.'[26]

These speculations by Fairholme, Cockburn, and others have since become part and parcel of the fundamentalist position regarding geology; in an updated form they occur in the later work of Price, Morris, and other American writers.[27]

---

[24] *Geology of Scripture*, 343. See also Fairholme, *Positions géologiques*, 1834.

[25] *The Age of the Earth*, 1838.

[26] 8.

[27] Price, *Evolutionary Geology and the New Catastrophism*, 1926. See Whitcomb and Morris, *Genesis Flood*, 1962. See also Barr, *Fundamentalism*, 1977.

# 17
# Cultural Impact of the New Perspective

Historical geology represented the avant-garde of science which had begun to roll back the frontiers of traditional subjects of academic study. Its perspective of earth history took the place of traditional, Mosaical cosmogony, and relegated the latter to the fringe of English culture. The vigour of the new geology was apparent in other ways; its individual discoveries and its overall perspective on time and history found their way into the language and literature of the period. This process of cultural absorption was aided by the fact that some of the discoveries of paleontology corresponded to existing popular beliefs. A resemblance was noticed between the extinct monster reptiles and the dragons of ancient mythology or medieval sagas. Similarly, monster stories in the Bible and the sea-serpents of contemporary folklore were brought to bear upon the paleontological discoveries, and vice versa.

Geologists themselves drew a comparison with mythology; Pidgeon referred to 'those innumerable reptiles, whose varied structure and colossal dimensions rival, if not surpass, the fabled monsters of poetical antiquity'.[1] But the popular literature, unconcerned with the time-scale of earth history, went further and hinted at an identity of mythical dragons and paleontological monsters. In Rennie's *Conversations on Geology* (1828, 1840) Edward says: 'Perhaps, then, the ancients were not so fabulous as we think them, in talking of their harpies, griffins, and dragons.' Mrs R. agrees and points out the resemblance between the dragon slain by St. George and a particular fossil skeleton.[2] The editor of the *Literary Gazette* suggested in a review of 'The fossil reptiles of England' (1841) that the fossil monsters 'seem to be the prototypes of fable, superstition, and romance'.[3] A correspondent suggested that the

[1] *Animal Kingdom*, 1830, 30.
[2] 1840, 304.
[3] 14 Aug. 1841, 513.

ornamental dragons of China might have their origin in paleontological reptiles: 'How curious it would be, if we should be looking
over the delineations of our China vases, teapots, etc., to improve
by a reference to five-clawed dragons, our ideas of the personal
appearance of Messieurs the Plesiosauri, Ichthyosauri, Iguanodons,
etc., etc.!'[4]

The dinosaurs suggested to many a new exegesis of such animal
references in the Bible as the *gedolim taninim* or 'great sea monsters'
of the creation story, and the behemoth and leviathan, best known
from their description in the book of Job. Thomas Hawkins inti-
mated the identity of paleontological monsters with biblical
examples in his books on Liassic reptiles (1834, 1840). However,
his bombastic style, his megalomania, and his querulousness
reduced the credibility of his books; a reviewer described one of
these as 'the largest jest-book we have ever seen'.[5]

Less idiosyncratic authors, however, made similar identifications. Thomas Thompson of the Hull Literary and Philosophical
Society argued that the behemoth and the leviathan correspond
with the megalosaur and the iguanodon. He believed that dinosaurs could have survived in the tropics until Old Testament times,
when Job would have been able to observe these monsters.
Thompson concluded: 'To sum up, then, I trust I have shown you
that there is good ground for supposing that the leviathan of the
Scriptures is the same animal as the now fossil megalosaurus; and
that the behemoth was identical with the iguanodon.'[6] This belief
received wide press coverage and was reported in both the *Magazine
of Natural History* (1835) and in the *Edinburgh New Philosophical Journal*
(1835). Among those who took the biblical and paleontological
monsters to be identical was Nares, Oxford's regius professor of
modern history.[7]

This identity implied that the monster reptiles had been contemporaneous with man and might still be alive today. Kirby speculated in his Bridgewater Treatise (1835) that animals like the
ichthyosaur and plesiosaur still inhabit hidden, subterranean seas;
'it would not be wonderful', he wrote, 'that some of the Saurian

[4] Ibid., 543.

[5] *Athenaeum*, 21 June 1834, 470.

[6] 'An Attempt to Ascertain the Animals Designated in the Scriptures by the Names
Leviathan and Behemoth', *Mag. Nat. Hist.* viii (1835), 320. See id., *Edinburgh New Phil.
Journ.* xix (1835), 263–81.

[7] *Man*, 1834, 185–6.

race, especially the marine ones, should have their station in the subterranean waters, which would sufficiently account for their never having been seen except in a fossil state.'[8] Most of Kirby's colleagues rejected this idea, but his notion that these animals might still exist was given support by sightings of sea-serpents in the Atlantic Ocean. Pontoppidan's *Natural History of Norway* (1755) had included a famous chapter on sea monsters, and in the first half of the nineteenth century a number of new sightings were reported. Two instances attracted international attention. The first involved observations of a sea-serpent off the coast of Massachusetts in 1817 and 1818, the second aboard HMS *Daedalus* in 1848, between St. Helena and the Cape. The credibility and interest of these sea-serpent stories were enhanced by the near-simultaneous discovery of the Liassic reptiles. Silliman's *Journal* (1820) gave extensive coverage to the sightings and published eye-witness accounts given under oath.[9] Silliman and Bakewell joined forces to suggest that the ichthyosaur might still exist and be the observed sea monster.[10] The sea-serpent became a legitimate topic for discussion. Duncan gave a talk on it to the Ashmolean Society (1833), and concluded that some rare and unknown sea monster did exist.[11]

A comprehensive discussion of the issue was published by Gosse in his *Romance of Natural History* (1860). He speculated that the sea-serpent was a form of plesiosaur, namely the very long-necked *Enaliosaurus*, or rather a large, modern descendant of it: 'I express my own confident persuasion, that there exists some oceanic animal of immense proportions, which has not yet been received into the category of scientific zoology; and my strong opinion, that it possesses close affinities with the fossil *Enaliosauria* of the lias.'[12] Such was the popular interest in large fossil monsters that an American, Albert Koch, forged a huge serpentine skeleton which he called *Hydrarchos Sillimani*, later changed, when Silliman objected, to *H. Harlani*.

These and similar instances show how readily the discoveries of

[8] *Animals and their History Habits and Instincts*, i (1835), 33–4.
[9] 'Documents and Remarks Respecting the Sea Serpent', *Am. Journ. Sci.* ii (1820), 147–64.
[10] See Bakewell, *Introduction*, 1838 edn., 362.
[11] 'On Sea Serpents', read 29 Nov. 1833, *Abstracts Proc. Ashmolean Soc.* i (1844), no. iii, 4–5.
[12] 368. See also Heuvelmans, *In the Wake of the Sea-Serpents*, 1968.

paleontology interlocked with popular beliefs, superstitions, and possibly fears. It facilitated the cultural absorption of the new science, a process demonstrated by the ease with which Dickens included an extinct fossil monster in the famous opening lines to his *Bleak House* (1852). Evoking the primeval atmosphere of London autumn weather, Dickens wrote: 'Implacable November weather. As much mud in the street, as if the waters had but newly retired from the face of the earth, and it would not be wonderful to meet a Megalosaurus, forty feet long or so, waddling like an elephantine lizard up Holborn Hill.'

The belief that fossil monsters from the Lias may have survived and be contemporaneous with man was partly based on an ignorance of the enormous length of the time-scale of earth history, which placed the 'age of the reptiles' long before the appearance of any humans. To some authors, such as Kirby, Nares, and probably also Gosse, the inferred contemporaneity of fossil monsters and man served as an argument against the scale of geological time. But to others the very length and complexity of geological history formed a source of literary inspiration. Geologists themselves, minor poets, and great names of English literature all, in varying degrees, used the new geology, its vocabulary, its images, and the substance of its theory.

SCAFE'S ALLEGORIES

In 1821 John Henry Newman wrote to his mother about Buckland's geology, saying that it was 'most entertaining, and opens an amazing field to imagination and to poetry'.[13] Students and colleagues alike wrote serious and humorous verse about Buckland's lectures on earth history, just ·as they did about his hyena den theory. Other subjects also were put into verse, from ammonites to the interior of Mantell's museum. Even Buckland's geological hammer was made the subject of an ode (1821) by Conybeare:

> Beneath the storm of its thundering blows
> > Rending, and opening, and staggering, and reeling,
> Mountains reluctant their story disclose,
> > The secret of millions of ages revealing.[14]

[13] 8 June 1821, *Letters and Diaries of Newman*, i, ed. Ker and Gornall, 1978, 109.
[14] Daubeny, *Fugitive Poems*, 78–9.

In 1823 *Ars Geologica* was chosen at Oxford as the subject of the chancellor's prize for Latin verse, an indication of the remarkably rapid rise of geology. The winning entry was by Isaac Williams of Trinity College, a friend of Keble. Williams later became embroiled in the Tractarian controversy when he failed to get elected to the professorship of poetry.

Around 1820 an intriguing story unfolded concerning some didactic verse on geology by a minor poet, John Scafe. He had been educated at Oxford, before Buckland's time, and then found employment in the army. In 1818 Scafe published a limited edition, twenty-five copies, of a poem entitled *King Coal's Levee*. It was an allegory which treated the geological succession of rocks and formations in the form of a court scene. At Oxford the geologists were delighted with Scafe's poem; Conybeare suggested improvements to the text and Buckland assisted with footnotes to a second edition of 1000 copies. Its popularity led to three new editions (1819, 1819, 1820), and his success motivated Scafe to similar verse, and in *Court News; or, the Peers of King Coal* (1820) he extended his allegory to include Buckland's table of the 'Order of Superposition of the British Strata'. He added a further collection which began with *A Geological Primer in Verse* (1820).

Part of the latter collection was reprinted in Silliman's *Journal* (1822) as 'at once specimens of skilful poetry, and of a lucid exhibition of geological facts and doctrines'.[15] *King Coal's Levee* became something of a sensation. In Germany Goethe summarized its contents and began (but never published) a German translation (1824).[16] Because of Buckland's and Conybeare's role, the poem appeared to be so closely associated with the university that Scafe's *Geological Primer* rhymed:

> A was an Agate as round as a Ball.
> B was Basalt in the cave of Fingal.
> C was King Coal, of Oxford the pride.[17]

The tale of *King Coal's Levee* ended, as does the fossil succession, with the Diluvium. In the allegory a rude rout of pebbles rush forward at the levee, but are ordered out:

[15] *Am. Journ. Sci.* v (1822), 272.
[16] *Goethe. Die Schriften zur Naturwissenschaft*, ed. Wolf. *et al.* xi (1970), 235–7.
[17] 21.

Plebeian PEBBLES, by odd BRECCIA cheer'd,
*Sans* loyalty, *sans* reverence appear'd.
Jostled, and rush'd, with frantic gabblings, on;
When the King roar'd, 'Gnomes, bid those brutes begone!
'Haste, drive them forth!—What? will ye stand like stone,
'And see your monarch bearded on his throne?'[18]

This treatment of the pebbles displeases their ruler, Giant Gravel,
who approaches adorned with all the fossil regalia associated with
diluvial deposits:

A monstrous sledge upon their vision burst,
Form'd of a mammoth's skeleton revers'd.
An elephant's bright tusks adorn'd its head;
Behind an elk's wide antlers backward spread.
It came,—as hippopotamus, huge, slow,
Rhinoceros, and heavy buffalo,
Yok'd in alternate pairs,—a fearful drove,—
With dull and measur'd trampling onward strove.
Within that sledge to brooding thoughts a prey,
Half stretch'd at length, the Giant GRAVEL lay.[19]

Giant Gravel then predicts a triumph of his pebbles over King Coal.
    The political atmosphere at the time was so charged that Scafe's
poem led to accusations of an Oxford Radical conspiracy. The
restoration of peace in 1815 had been followed by economic and
social distress which fostered Radical politics. The government
passed the Gagging Acts of 1817 which restricted meetings;
penalties were increased for treasonable and seditious propaganda.
The establishment's anxiety about Radical subversion was such
that *King Coal's Levee* was seen by some people as a political allegory
with seditious implications. Who but a Radical would use coal as
the symbol of the monarch? Who else but the common Radical
rabble could be meant by the pebbles? Clearly, their predicted
victory over King Coal constituted a form of revolutionary incite-
ment. J. J. Conybeare, in his dual capacity as geologist and
professor of poetry, had to allay the fears of an Oxonian conspiracy

[18] 35.
[19] 37–8.

and Scafe had to add a disclaimer to his poem assuring the reader of no other than geological intent.[20]

## TENNYSON'S POETRY

Scafe wrote his allegories when the study of earth history was still in its adolescence, several years before the theory of progressive development had become a major issue. This theory was an intellectual novelty when Alfred Tennyson was a student at Cambridge (1827–31). Tennyson displayed a keen interest in the advances of science, not least in those of geology. He was deeply disturbed by the premature death of his close university friend Arthur Hallam in 1833. Personal distress and social issues formed the inspiration for his two great poems, begun at different times during the 1830s, but finished much later, *The Princess* (1847) and *In Memoriam* (1850) (of Arthur Hallam).

In these poems Tennyson's anxiety about the apparent transience of life was expressed in an accurate and skilful use of the desolate perspective of earth history, its successive worlds and its episodic extinctions. His mastery of both the detail and the theory of progressivism and the expressive beauty of the passages in which he described these were such that Tennyson may be seen as the poetical exponent of the English school of geology.

This conclusion is at variance with the standard interpretation of the geology in Tennyson's 1840s poetry. It has become customary to read in *The Princess* and in *In Memoriam* not only organic evolution, but also the geology of such opponents of the English school as Lyell. Because the relevant sections were written not only long before Darwin's *Origin of Species*, but even before Chambers's *Vestiges*, some literary critics have interpreted these passages as an anticipation of the theory of organic evolution by the intuitive genius of a poet, before the analytical mind of Darwin dared to arrive at the same conclusion. 'How did a poet come to forestall the scientists in their own game?', one critic asks.[21] A considerable body of literature has accumulated on both sides of the Atlantic asserting

[20] (Scafe), *A Critical Dissertation*, printed with a *Geological Primer in Verse*, 1820, 49–67. See also Buckland to Mary Cole, 13 July 1819, 29 Oct. 1819, 18 Sept. 1821, NMW, Bu P; Buckland to Greenough, 26 Mar. 1819, 7 June 1819, 25 Aug. 1821, CUL, Gr P; Boase and Courtney, *Bibliotheca Cornubiensis*, ii (1878), 627. Scafe's other poetry was published under the title *The Genius*, 1819.

[21] Stevenson, *Darwin among the Poets*, 1932, 55.

Tennyson's supposed anticipation of organic evolution and the Lyellian derivation of his geology.[22]

Not only is this interpretation of Tennyson's science entirely wrong, but the very opposite is true. Neither organic evolution nor Lyellian geology occurs in *The Princess* or in *In Memoriam*. The reading of organic evolution into these poems is based on the elementary mistake of confusing the theory of progressive succession with that of evolution; progressivism was included by Chambers in his *Vestiges*, but he added species mutability. No trace of the latter occurs in the 1840s poems. Tennyson is known to have said that the *Vestiges* contained speculation which he himself had written about in more than one poem; as the bulk of Chambers's book dealt with progressivism as distinct from species change there is no basis in this statement for an evolutionary interpretation of Tennyson's paleontology.

The raw material for the Lyell industry in Tennyson criticism is to be found in a single sentence written by Hallam Tennyson in the 'Life and Letters' *Memoir* (1897) of his father: 'During some months of 1837 my father was deeply immersed in Pringle's *Travels,* and Lyell's *Geology.*'[23] There is no reason to doubt that Tennyson read Lyell; so did many other people, including George Eliot. But to conclude that, *eo ipso*, Lyell was the source of Tennyson's geology, is feeble.

At the time when Tennyson's son wrote the *Memoir* it was considered more enlightened to have read Lyell than such clerics as Buckland and Sedgwick. This was not so during the 1830s, and Tennyson must have been familiar with the geology of the English school. Cambridge was one of its centres, and during his years of residence there his own tutor Whewell was involved in the formulation of the anti-Lyellian, progressivist synthesis of earth history. At Cambridge too Tennyson attended Sedgwick's lectures and, with Arthur Hallam, followed the discoveries in paleontology. The latter wrote to Tennyson in 1832: 'I should feel like a melancholy Pterodactyle winging his lonely flight among the linnets, eagles, and flying fishes of our degenerate post-Adamic world.'[24] This

[22] E.g. Culler, *Poetry of Tennyson*, 1977, 150; Turner, *Tennyson*, 1976, 108–9; *Tennyson. In Memoriam*, ed. Hunt, 1970, *passim*. Even Ricks's edition of the *Poems of Tennyson* perpetuates these myths (1969), 910–11. All quotations are from Ricks's edition.
[23] i (1897), 162.
[24] Ibid., 85.

effortless use of a paleontological metaphor came not long after Buckland had introduced the pterodactyl into the English language. A later Cambridge man, Richard Wilton, believed that Tennyson had not only used the substance of Sedgwick's lectures, but even his language; he wrote: 'His lectures are a rich mine of strong, rugged, and picturesque English; and I am confident Tennyson has worked in it assiduously. I could quote many passages to prove that he has studied and imitated Sedgwick's grand, nervous style.'[25]

Apart from these circumstantial considerations the content of Tennyson's poems speaks with unambiguous clarity the language of the English school. Its distinctive features are all present, namely an interest in vertebrate fossils (especially giant reptiles), an emphasis on the finite duration of species shown by episodic extinctions, a reluctant acceptance of the nebular hypothesis with its corollary of a central heat, and a committed belief in progressive change. These sharply contrasted with the views of Lyell, who reiterated, at about the time Tennyson finished *In Memoriam,* his opposition to progressive succession in a widely publicized address to the Geological Society.[26]

Vertebrate paleontology and extinction are combined in *The Epic* (1842): 'For nature brings not back the Mastodon.'[27] The excitement generated by the discovery of reptilian foot prints such as those of the *Chirotherium* found its way into *The Princess*:

> '. . . Would, indeed, we had been,
> In lieu of many mortal flies, a race
> Of giants living, each, a thousand years,
> That we might see our own work out, and watch
> The sandy footprint harden into stone.'[28]

The brevity of human existence is expressed by an effective use of vertebrate fossils:

> She bowed as if to veil a noble tear;
> And up we came to where the river sloped
> To plunge in cataract, shattering on black blocks
> A breath of thunder. O'er it shook the woods,

[25] Wilton to Morine, 6 Nov. 1849, SM, SelP.
[26] It was published in the *Edinburgh New Phil. Journ.* li (1851), 1–31, 213–26.
[27] 584.
[28] 780.

> And danced the colour, and, below, stuck out
> The bones of some vast bulk that lived and roared
> Before man was.[29]

In the same poem the progressivist synthesis is summarized in fewer than eight lines:

> 'This world was once a fluid haze of light
> Till towards the centre set the starry tides.
> And eddied into suns, that wheeling cast
> The planets: then the monster, then the man;
> Tattooed or woaded, winter-clad in skins,
> Raw from the prime, and crushing down his mate;
> As yet we find in barbarous isles, and here
> Among the lowest.'[30]

This mention of the nebular hypothesis was followed by a reference to a central heat in *In Memoriam*; it was expressed with a characteristic lack of commitment:

> They say,
> The solid earth whereon we tread
>
> In tracts of fluent heat began,
>     And grew to seeming-random forms,
>     The seeming prey of cyclic storms,
> Till at the last arose the man;[31]

Tennyson was both fascinated and horrified by the cruel and merciless violence attributed to the ancient monsters. De la Beche had so portayed them in his *Duria Antiquior,* and Martin had visualized the reptilian monsters of the geological past as locked in mortal combat, e.g. in his 'The country of the Iguanodon' for Mantell's *Wonders of Geology.* To such scenes Tennyson's famous lines from *In Memoriam* refer: 'Nature, red in tooth and claw', or 'Dragons of the prime, that tare each other in their slime'. No anticipation of a Darwinian 'survival of the fittest' should be read into these lines.

> 'So careful of the type?' but no.
>     From scarped cliff and quarried stone
>     She cries, 'A thousand types are gone:
> I care for nothing, all shall go.

[29] 781.
[30] 762.
[31] 969.

'Thou makest thine appeal to me:
I bring to life, I bring to death:
The spirit does but mean the breath:
I know no more'. And he, shall he,

Man, her last work, who seemed so fair,
Such splendid purpose in his eyes,
Who rolled the psalm to wintry skies,
Who built him fanes of fruitless prayer,

Who trusted God was love indeed
And love Creation's final law—
Though Nature, red in tooth and claw
With ravine, shrieked against his creed—

Who loved, who suffered countless ills,
Who battled for the True, the Just,
Be blown about the desert dust,
Or sealed within the iron hills?

No more? A monster then, a dream,
A discord. Dragons of the prime,
That tare each other in their slime,
Were mellow music matched with him.

O life as futile, then, as frail!
O for thy voice to soothe and bless!
What hope of answer, or redress?
Behind the veil, behind the veil.[32]

The sense of futility and despair expressed in these moving stanzas was instilled by the new perspective of earth history and by its implications for the future. Man's position in nature appeared not to be unique, but to be no more than a single link in the great chain of history, preceded by many worlds which had become extinct, to be succeeded by new worlds possibly superior to his own. One day man would exist no more and his remnants be fossilized 'or sealed within the iron hills'. This prospect dimmed the Christian hope of redemption, of resurrection and a future life of eternal beatitude on a new earth, governed by love, and free from primeval violence. The very reason why many traditional believers objected to the geology of the English school and why Whewell

[32] 911–12.

wrote his *Plurality of Worlds* was the reason why Tennyson saw no 'hope of answer, or redress'.

The new geology was used by several other artistic and literary figures. Ruskin was one of Buckland's protégés during his student days at Christ Church when he prepared illustrations for the geology lectures. In later life Ruskin kept in touch with the Buckland family, especially with Mary, showing an imaginative though somewhat idiosyncratic interest in geology.[33] Another Christ Church man, Charles Dodgson (Lewis Carroll), used one of the show-pieces of Oxford geology, the dodo, in his *Alice's Adventures in Wonderland* (1865). English geology was in most instances indirectly transmitted to continental literature, e.g. by Humboldt's *Kosmos* (1845–62). Victor von Scheffel's *Gaudeamus* (1869) contained a variety of poems on geological subjects and, best of all, a satirical poem 'Der Ichthyosaurus' (written 1854) which included the various monster reptiles, even a coprolite, and combined good verse with fine satire which gently mocked the moral interpretation which natural theology had given to earth history.[34]

---

[33] See for example Ruskin to Mary Buckland, 10 Feb. 1856, DRO, Bu P.

[34] *Scheffels Werke,* ed. Panzer, i (1919), 14–15. See also the English translation by Leland, *Gaudeamus!,* 1872. Buckland's Bridgewater Treatise may have also exerted a direct influence abroad. Two German translations came out (1837, 1838–9), and two French (both 1838). One of the French editions was very successful. Buckland wrote to Brougham, 1 May 1839, UC, Br P.: 'I have just heard from Paris that the French Academy of Sciences has awarded a prize of 3,000 franks to the translator of my Bridgewater Treatise, as being the most useful and important work pour la morale that has appeared in France during the last year. And further that it is appointed as one of the annual Prize Books to be distributed in the provincial colleges of France.'

# PART III

# PROVIDENCE IN EARTH HISTORY

## *The Divine Right of Geology and of Political Economy*

>                                             They say,
> The solid earth whereon we tread
>
> In tracts of fluent heat began,
>     And grew to seeming-random forms,
>     The seeming prey of cyclic storms,
> Till at the last arose the man;
>
> Who throve and branched from clime to clime,
>     The herald of a higher race,
>     And of himself in higher place,
> If so he type this work of time
>
> Within himself, from more to more;
>     Or, crowned with attributes of woe
>     Like glories, move his course, and show
> That life is not as idle ore,
>
> But iron dug from central gloom,
>     And heated hot with burning fears,
>     And dipt in baths of hissing tears,
> And battered with the shocks of doom
>
> To shape and use. . . .
>
> Tennyson, *In Memoriam A. H. H.*, cxviii. 7–25

# 18
# The Natural Theology of Fossils

## A VEHICLE FOR GEOLOGY

The argument from design states that order in nature indicates the existence of a supreme designer or creator. This argument for divine existence was discussed in classical and medieval philosophy and gained new popularity in early modern times. Copernicus and Kepler wrote about the elegance of the cosmic design. Boyle developed the analogy between a clock and its maker and the universe and its creator, when the new mechanical philosophy was coming into vogue. In the early eighteenth century Derham contributed further examples of design from natural history, and Newton did the same from cosmology. The design argument, represented by what became known as natural theology or physico-theology, was a favourite way of refuting atheism. Bernard Nieuwentyt's *Regt Gebruik der Werelt Beschouwingen* (1715) was a particularly fine example of this. The argument by analogy became very popular in England, partly as a result of Butler's *Analogy of Religion, Natural and Revealed* (1736).

The design argument was criticized by Hume in his *Dialogues concerning Natural Religion* (posthumously 1779), e.g. by a *reductio ad absurdum* of the reasoning by analogy: if we consider the perfection of nature as implying a designer, then, by parity of reasoning, that perfect designer must imply a prior designer, and so on *ad infinitum*.[1] In traditional histories of ideas the decline of the argument from design is held to begin with Hume and is traced past Kant's scepticism down to rock bottom with Darwin. The latter turned the argument inside out by formulating his mechanism of natural selection which states that, given variability in nature, anything that is not adapted, i.e. not well designed, will simply not survive.[2]

This decline curve is inaccurate for England and even inapplicable

[1] *Hume's Dialogues concerning Natural Religion,* ed. Smith, 2nd edn 1947, 161 and *passim.*
[2] *Dictionary of the History of Ideas,* ed. Wiener, s.v. 'Design Argument'.

to Scotland. The heyday of natural theology with its argument from design was still to come in the course of the early nineteenth century. A new wave of the tradition's popularity began with Paley's *Natural Theology* (1802) and crested during the 1830s when the Bridgewater Treatises appeared. These derived their name and origin from Francis H. Egerton, the eighth earl of Bridgewater, who combined an interest in manuscript collecting, dogs, rabbits, and illegitimate daughters with a concern for design and purpose in the world. In 1825 he made a will which left £8,000 in trust to the president of the Royal Society, who was to select and reward a person or persons to publish on the subject of the 'Power, Wisdom and Goodness of God, as manifested in the Creation'. When Egerton died in 1829 Gilbert Davies, the then president of the Royal Society, solicited the help of the archbishop of Canterbury (William Howley) and the bishop of London (Charles Blomfield) in choosing eight authors for the execution of the earl's will. The resulting treatises (1833–6) were necessarily uneven in quality, but they covered almost the entire spectrum of contemporary science, including such recently formalized subjects as political economy and geology.

The series was enlarged by uninvited contributions, e.g. Frederick Bakewell's *Natural Evidence of a Future Life* (1836) and Charles Babbage's *Ninth Bridgewater Treatise* (1837). A further high point of natural theology in the 1830s was the new edition of Paley's classic (1835–9), edited by Bell, one of the Bridgewater Treatise authors, and Brougham, who himself wrote a separate introductory *Discourse on Natural Theology*.

Natural theology was a decidedly pluralistic cultural phenomenon. The only common denominator among its advocates was opposition to atheism. Opinions on many issues differed and natural theology was used for a variety of purposes. At Oxford and Cambridge it formed a guise under which science claimed the right to an academic niche. In the country at large natural theology was used to give moral justification to particular economic policies and social theories. The generality of the argument from design made it suitable for interdenominational co-operation in the furtherance of science, exemplified by the British Association.[3] However, long before Darwinism undercut the argument, its latitudinarian

[3] See Brooke, 'The Natural Theology of the Geologists', *Images of the Earth*, ed. Jordonova and Porter, 1979, 39–64.

character formed a major reason why it was mistrusted by tradition-alists, especially by the Oxford Tractarians.

Several leading members of the English school were active participants in the natural theology of the early nineteenth century. In some respects geology, and especially paleontology, occupied centre stage among the various subjects of natural history, from which arguments were extracted to prove 'the power, wisdom, and goodness of God'. It has been said that English geology contributed so substantially to the literature on natural theology in order to show that it did not have an inherent tendency toward irreligion or philo-sophical scepticism.[4] The traditional, self-professed purpose of geology in its role of natural theology was indeed to refute atheism and other forms of irreligious belief. There was, however, a more down-to-earth reason for geology presenting itself as a new source of material for the argument from design.

The purpose of refuting atheists was largely an echo from the eighteenth century. In the early nineteenth century, at the old English universities, even Hume's scepticism was no longer a major issue; to present new arguments against atheists and other free-thinkers was to attack opponents of straw. Whately's clever booklet, *Historic Doubts relative to Napoleon Buonaparte* (1819), purporting to refute Hume's argument against miracles, was an affirmation of the utility of historical evidence. Its many new editions and its translations were readily addressed to contemporary issues. The German edition (1836), for example, used it to counter Strauss's *Das Leben Jesu*.

A live issue was educational reform; the improvement of academic standards by new examination statutes and the modern-ization of the old system of classical education, by providing new subjects, were of more immediate concern than the refutation of atheistical systems of belief. Oxford and Cambridge needed an apologia against the accusations in the *Edinburgh Review* of academic backwardness, not against Hume's philosophical scepticism.

Under these pressures of curricular change and reform, geology's presentation of itself as a form of natural theology aided its introduction as a new subject of academic study. Although geology was novel, its contributions to the argument from design demonstrated its usefulness to, and compatibility with, the existing

---

[4] E.g. Raven, *Natural Religion and Christian Theology*, 1953, 173; Chadwick, *Victorian Church*, i (1971), 560.

tradition of learning. In order to justify teaching geology and to find financial security for its professors it had to be proved useful as a hand-maiden to theology, or, better still, appear to be a form of theology itself. The pressure on geology at Oxford and Cambridge to acquire ecclesiastical approval and patronage conditioned it to use the guise of natural theology; it was thus constrained to a far greater extent at these universities than at continental ones, at Edinburgh, or, later on, at the university of London, where geology's worth was not judged by its contribution to clerical education. Geology in its guise of natural theology did function as a form of religious apologetics in opposition to atheism, but the primary reason why English geology contributed so explicitly and substantially to the argument from design was its Oxbridge habitat. There its self-presentation as natural theology was not so much a halo of religious purity, but a hood of academic accreditation demonstrating its usefulness as a subject of university education.

Very few prominent geologists on the Continent or in Scotland, irrespective of their religious commitment, contributed to the literature on natural theology. The English school was unique in emphasizing rocks and fossils as a basis for the argument from design. Neither Cuvier nor Humboldt did anything similar; Agassiz explicitly dissociated himself from Buckland's natural theology;[5] in Edinburgh Jameson added little to the design argument, and Lyell, who resented Buckland's ecclesiastical patronage, rarely concerned himself with this type of literature.

Among the members of the English school, the Oxford and Cambridge representatives were the main contributors to natural theology: Buckland, Conybeare, Sedgwick, and Whewell. Both Buckland and Whewell were Bridgewater Treatise authors and their books were probably the best of the entire series. Outsiders, by no means irreligious, such as de la Beche, Mantell, Murchison, and Phillips, made comparatively few references to natural theology. Other subjects of natural history, anatomy and botany, also had to make an effort to attract a lecture audience at the ancient English universities, and although they were under no suspicion of irreligion they too presented themselves explicitly and vigorously as forms of natural theology.

During successive decades the contributions of geology to natural theology changed. Initially, in particular during the 1820s,

[5] *Geologie und Mineralogie*, i (1839), 124

geology as diluvialism contributed not only to natural, but even to revealed, religion, apparently corroborating the story of the Mosaic deluge. When around 1830 this corroboration was withdrawn and the English school developed its progressivist synthesis of earth history, much emphasis was given to features of functional adaptation and design. A further change of emphasis occurred when the geological argument from design moved its searchlight away from manifestations of God's existence towards its implications for human society. It thus became embroiled in issues of moral philosophy and political economy about which Buckland and his Cambridge colleagues disagreed.

It is commonly supposed that Paley's *Natural Theology* contributed to the revival of science at Oxford and Cambridge during the early nineteenth century and to the cultivation of a taste for natural history in the country at large. Paley's name and his illustrations of design were indeed extensively used at Oxford to help establish new science courses. The university's scientists argued that their subjects were academically useful because they contributed to natural theology. Buckland argued this for geology, Kidd for anatomy, Daubeny for botany, and the Duncan brothers for natural history in general and for museum exhibitions in particular.

Significantly, this use of natural theology was rarely made in professional papers presented to colleagues at meetings of scientific societies, but often in inaugural lectures and sermons directed at the university audience at large. Such public presentations went hand in hand with a romanticized appeal to the halcyon days of experimental science of the middle-to-late seventeenth century, when men such as Boyle, Newton, and Ray had made the old English universities the cradle of scientific discoveries and enlisted these in the cause of natural theology. The argument from design had become institutionalized in the Boyle lectures, in which series Derham had given his classic sermons on *Physico-Theology* (1711–12).

The very title of Buckland's inaugural lecture of 1819, *Vindiciae Geologicae,* carried a Miltonian connotation of natural theology 'to justify the ways of God to man' by means of geological phenomena. However, such justification of God from geology amounted to a vindication of geology to the university. Buckland argued that, even though his subject was of economic utility, its entitlement to a place at Oxford rested on its contributions to physical and moral truth. In this respect geology ranked with the works of Newton and Paley. It not only provided proofs of the existence of God and of his attributes but also indicated his unity by showing a unity of anatomical plan among fossils from successive geological worlds.[6]

Kidd was equally explicit in his use of natural theology to promote science at Oxford. The title of his inaugural lecture on his appointment to the regius chair of medicine included Paley's name: *An Introductory Lecture to a Course in Comparative Anatomy, illustrative of Paley's Natural Theology* (1824). Kidd had done more than anyone else during the first two decades of the nineteenth century to kindle an interest in science at Oxford. In the preface to his lecture he expressed some satisfaction: 'Oxford is now strong enough to say without boasting, that though clouds soon obscured her bright dawn of natural science, which, originating in the labours and association of Boyle, Hooke, Wallis, Sir Christopher Wren, and others of similar pursuits, subsequently led to the establishment of the Royal Society of England, that dawn has again revived.'[7] Kidd made a strong plea for the further improvement of natural history education at Oxford organized along the lines of Paley's exemplary design argument. Thus his own anatomy lectures would not only be of interest as a natural science but also have a direct bearing on theology.

The same use of natural theology was made by the Duncan brothers. When P. B. Duncan published the *Catalogue to the Ashmolean Museum* (1836) he wrote about the middle 1820s: 'Happily at this time a taste for the study of natural history had been excited in the university by Dr. Paley's very interesting work on Natural Theology, and the very popular lectures of Dr. Kidd on Comparative Anatomy, and Dr. Buckland on Geology.'[8] He mentioned that his brother had enhanced the interest of the exhibition of the specimens by using Paley, giving 'an exalted interest to the collection, such as no exhibition of the kind has hitherto displayed'.[9] He did not elucidate what was meant by this Paleyan arrangement, but from J. S. Duncan's *Botano-Theology* (1825), and more particularly his *Analogies of Organized Beings* (1831), it would appear

[6] 13–14.
[7] vii–viii.
[8] vi.
[9] Ibid.

that specimens were not just arranged in a taxonomic order, but in groups which showed both homologies of identical anatomical elements adapted to different environments (the fin of a fish and the wing of a bird) and analogies of functional adaptation to a single environment by different anatomical elements (the wing of an insect and the wing of a bird).

When in the course of the 1830s the popularity of science declined at Oxford, after its auspicious rise during the preceding decade, the plea for reform and a provision for new subjects became more aggressive. Powell complained bitterly about the declining interest in scientific studies in his public lecture on the *Present State and Future Prospects of Mathematical and Physical Studies in the University of Oxford* (1832). Daubeny spoke even more bluntly in his inaugural lecture on *The Study of Botany* (1834). He used the argument formulated by his colleagues in previous inaugurals that science, and in this particular instance botany, was not only of practical utility but contributed to an education in theology, 'affording to the Divine some of the most beautiful illustrations of design with which nature can supply him'.[10] Daubeny went further, arguing that if the clergy wanted to retain control over Oxford education they must extend their patronage more actively to new subjects of scientific interest.

*Mutatis mutandis,* natural history as a source of design features was also used to promote science at Cambridge. Sedgwick's *Discourse* linked geology to the university's curriculum by emphasizing the contributions which the study of the earth made to natural theology. Conybeare, who had taken the ideal of his Oxford education to the West Country, drew a similar connection in his *Inaugural Address* (1831) to a course of theology lectures at the recently established Bristol College. He argued that a proper theology education not only required a study of the classics, but also of the sciences and their contributions to the argument from design.

Paley had made the human body the main source of evidence for design, though he also used plants and animals. Geology, however, provided a whole new range of phenomena from which adaptation and design could be inferred. Paleontology, especially, with its extinct and unfamiliar forms of life, enriched the standard examples of design in nature by adding new, and in some instances bizarre, contrivances from the geological past. Buckland's expertise as a paleontologist, combined with his ecclesiastical patronage, made

[10] 27. See also Powell, *Connexion of Natural and Divine Truth,* 1838.

him the obvious choice as author for a Bridgewater Treatise on geology and mineralogy. Buckland tried to make his book accessible to a broad readership, but he also wanted to write an up-to-date and authoritative summary of historical geology. He achieved this double aim by adding to a lively text a separate volume with plates. He enlisted the assistance of highly skilled colleagues—Agassiz, Broderip, Clift, Owen, Murchison, and many more. As a result his Bridgewater Treatise was delayed and came out last in the series. With pride and relief Buckland wrote to Featherstonhaugh in 1836: 'Lockhart the editor of the Quarterly, who is no naturalist or geologist, told me that he thought the book readable by all literary persons who can follow a logical train of argument, and Conybeare who has just seen it, is full of admiration and has requested to write another review of it immediately. He is surprised at the quantity of new matter which has caused the long delay.'[11]

## GOD'S OWN MEGATHERIUM

Buckland's most effective example of design was the structure of sloths, both living and extinct representatives. He chose this example deliberately, because famous continental naturalists, including Buffon and Cuvier, had singled out sloths as rare but real instances of poor design, animals which were imperfectly organized and condemned to live an abject life of inconvenience and misery. The anomalously long arms of sloths, for example, appeared by comparison with those of other quadrupeds as monstrosities, unfit for moving on all fours. Buckland took issue with this view. In a paper to the Linnean Society 'On the Adaptation of the Structure of the Sloths to their peculiar Mode of Life' (read 1833), Buckland argued that design and the degree of perfection of an animal's organization are a function of adaptation to a particular environment; fish can hardly be judged as imperfectly organized because they possess fins instead of legs. Once the sloths are placed in the context of their habitat, i.e. in trees, what seemed debility of anatomy proves to be perfect adaptation and thus perfect design: 'The extraordinary length of the arm and fore-arm, so inconvenient for moving on the earth, are of essential and obvious utility to a creature whose body is of too great weight to allow it to crawl to the

[11] 18 Apr. 1836, CUL, Se P.

extremity of the branches to collect the extreme buds and youngest leaves, which form its food: these long arms in fact perform the office of the instrument called "lazy tongs", whereby the creature brings food to the mouth from a distant point without any movement of the trunk.'[12] This was a valuable improvement on Cuvier's anatomy inspired by Paleyan natural theology. The design argument did not allow for arbitrary imperfections in nature.

As so often Buckland's choice of example and manner of presentation inspired caricature and jest. J. S. Duncan composed a poem on 'The Bradypus or Sloth' in which a roebuck passes under a sloth hanging from a tree and curses it as a 'misshaped abortion', 'unfit to run, to wade, to fly, to swim'; at that moment a serpent slings itself around the roebuck in an attempt to crush it, but the sloth grasps the serpent with its powerful claws and saves its victim, adding 'in Quaker strain':

> The pow'r which made thee light to scour the plain,
> Which arm'd the couguar's and the cayman's jaws,
> Which gave the condur wings, and beak, and claws,
> Which arm'd the bull with horns for furious fight,
> And man with craft which baffles bestial might,
> Gave to the birds *his* air, the fish *his* sea,
> And fits the Bradypus to grasp the tree:
> And in these arms, which mov'd thy scoff and scorn,
> Combines the pow'r of talon, tusk, and horn.[13]

The same reasoning was used by Buckland to interpret the skeleton of the megatherium (figure 13). This was one of several genera of giant sloths, all extinct, described by Cuvier as early as 1795, and subsequently by Pander and d'Alton in their monograph on *Das Riesen-Faultier* (1821).[14] Both these descriptions were based on a skeleton found near Buenos Aires in 1789 and exhibited in Madrid. In 1832 Woodbine Parish shipped another skeleton to Europe, destined for the Royal College of Surgeons, where Clift described it. A decade later Owen produced a systematic study of the various known megatheroid quadrupeds in his *Description of the*

---

[12] *Trans. Linnean Soc.* xvii (1837), 19. See also Burchell to Buckland, 13 Aug. 1832, DRO, Bu P.

[13] Daubeny, *Fugitive Poems*, 1869, 153–6.

[14] Pander and d'Alton, *Die vergleichende Osteologie*, i (1821–7).

*Skeleton of an Extinct Gigantic Sloth* (1842).[15] To these studies of anatomy and taxonomy Buckland contributed his interpretation of functional anatomy. The choice of the megatherium was both daring and effective; its huge frame, as large as that of an elephant, looked as awkward as that of the modern *Bradypus,* if not more so. Pander and d'Alton described the megatherium as ill-proportioned and misshapen, making the elephant appear slenderly built, the rhinoceros elegant, and even the hippopotamus wellshaped. 'I select the Megatherium,' Buckland declared, 'because it affords an example of most extraordinary deviations, and of egregious apparent monstrosity.'[16]

Buckland's interpretation of the functional anatomy of the megatherium became a *cause célèbre* of the natural theology of the 1830s. His skilful deduction of the animal's mode of living and habitat from its anatomy earned him universal admiration and applause. Buckland made it the subject of a lecture which ended the meeting of the British Association at Oxford in 1832. The composition of his audience symbolized the latitudinarian character of both the British Association and of natural theology: Anglican and dissenter, Tory and Whig, Oxbridge scholar and provincial naturalist, students of pure and of applied science, male and female, all mingled in the overcrowded Holywell Music Room on the evening of 23 June 1832.

Buckland began his lecture with a eulogy on Cuvier who had recently died. But then he criticized Cuvier's interpretation of the megatheroid skeletons as ill-designed. From a careful examination, particularly of the teeth and feet, Buckland made out a convincing case that these extinct giant sloths had been root-eaters and had lived on the South American pampas where they had to dig their food out of the stubborn ground. It were this mode of living and this habitat to which the megatheroid skelteon, however awkward in appearance, had been so finely adjusted. The following is a specimen of Buckland's lively discourse:

Gentlemen, his teeth indicated a pecularity of structure; they were not calculated to eat leaves or grass; they were not calculated to eat flesh; he was an eater of vegetables. What then remained for him but roots? He has a

---

[15] Owen corrected Buckland's mistake of having attributed the armour of the *Glyptodon* to the megatherium. See also Owen, 'Notice of the Glyptodon', *Proc. Geol. Soc.* iii (1842), 112–13.

[16] *Geology and Mineralogy,* i (1836), 144.

13. Skeleton of a megatherium, eight feet tall and twelve feet long, discovered near Buenos Aires in 1789.

spade, and he has a hoe and a shovel in those three claws in his right hand. . . . He is the Prince of sappers and miners—I speak in the presence of Mr Brunel the Prince of Diggers. Mr Brunel eyes him and says 'I should like to employ him in my tunnel.' 'No,' say I, 'he is not a workman for you; he is not a tunneller, he is a canal digger if you please, so I pray give him the first job you have to do.' He will not go an inch below a foot and a half; he would dig a famous gutter, he would drain all Lincolnshire in the ordinary process of digging for his daily food.[17]

Buckland called the giant sloth 'Old Scratch', and he interspersed his lecture with jests focused on the ladies in the audience, on various luminaries, and on issues of the day. By all accounts it was a remarkable performance, memorable to those present, and deservedly given a wide and favourable press coverage. Indeed Buckland's talk on 'Old Scratch' in the Holywell Music Room represented the apex of the nineteenth-century design argument in England. Mantell, himself an accomplished paleontologist, regarded Buckland's lecture as the grand finale of the week-long British Association meeting; he wrote in his diary: 'In the evening went to the Music room and heard the lecture of Dr. B. on the Megatherium! A very admirable discourse: the room crowded to excess: terminated at 12 o'clock at night, and thus closed the Session after a week of the highest possible intellectual enjoyment! It amply rewarded me for all the labor which scientific research has ever cost me. To Dr. Buckland for his liberality and generous attention I can never be sufficiently grateful!'[18] *Fraser's Magazine*, in its account of the British Association meeting, devoted a disproportionately large part to Buckland's closing talk; however, even though it approved of its content, the magazine disliked Buckland's jocular manner of presentation, which it believed to be 'blamable, if not disgusting, in such a place, and at such an institution as Oxford'.[19]

Buckland's lecture was never published. But much of its substance was included in his Bridgewater Treatise, which reiterated the functional interpretation of the megatheroid skeletons, though in a more sober manner. Thus the very awkwardness of the megatheroid anatomy turned the animals into the best-known

[17] 'Lecture Delivered by Buckland . . .', manuscript copy of Buckland's speech, 50-1, DRO, Bu P. See also *Report BAAS, 1831, 1832,* 104-7.
[18] *Journal of Mantell,* ed. Curwen, 1940, 104.
[19] v (1832), 735.

examples of perfect organization among vertebrate fossils. From the invertebrates Buckland selected the trilobites, known from the ancient Transition rocks, as his prime example of design. He paid particular attention to their eyes, which are composed of a delicate array of many tiny crystals. Buckland gave accurate illustrations of these and drew a comparison with similar eyes in modern crustaceous animals and in insects.

This analogy with the visual apparatus of living creatures supported Buckland's conclusion that a normal solar source of light must have been available during the early part of geological history when trilobites lived; both the composition of the atmosphere and that of the water in which they swam must have been sufficiently transparent to allow their eyes to function. The interest of this deduction depended not just on its geological significance; it also confirmed Buckland's exegesis of the first chapter of Genesis. He interpreted the creation of the sun, moon, and stars not as a *creatio ex nihilo*, but as an atmospheric reappearance of their light in the service of man, even though they had existed as celestial bodies from the very beginning of the universe.

The trilobite eyes provided Buckland with an example of design in the purest sense of the argument by analogy, made all the more remarkable by the age of the fossils:

If we should discover a microscope, or telescope, in the hand of an Egyptian Mummy, or beneath the ruins of Herculaneum, it would be impossible to deny that a knowledge of the principles of Optics existed in the mind by which such an instrument had been contrived. The same inference follows, but with cumulative force, when we see nearly four hundred microscopic lenses set side by side, in the compound eye of a fossil Trilobite; and the weight of the argument is multiplied a thousand fold, when we look to the infinite variety of adaptations by which similar instruments have been modified, through endless genera and species, from the long-lost Trilobites, of the Transition strata, through the extinct Crustaceans of the Secondary and Tertiary formations, and thence onwards throughout existing Crustaceans, and the countless host of living Insects.[20]

Even those who had not always been in agreement with the diluvialism or progressivism of Buckland and the English school could admire the natural theology of *Geology and Mineralogy*. Scrope, in his review of the book for the *Quarterly Review*, noted that the use of paleontological specimens in the context of the familiar and popular

[20] *Geology and Mineralogy*, i. 403.

argument from design facilitated the acquaintance with geology by the public at large and 'that Dr. Buckland will be the means of introducing many a saurian, many a trilobite, and many an encrinite to the acquaintance of those who would hardly have heard of such beings but for his excellent book'.[21]

In an earlier review of the *Principles of Geology* Scrope had acknowledged Buckland's Bridgewater Treatise in anticipation of its appearance. He regarded Lyell's three-volume work as a fit introduction to Buckland's book in matters of natural theology.[22] Even though Lyell himself made few if any references to natural theology, he praised Buckland for his contributions to the design argument. In his anniversary address of 1837 Lyell commented: 'It is impossible to read the account given of the Megatherium, and to contrast it with that drawn up by Cuvier of the same species, without being struck with the increased interest and instruction, and the vast accession of power derived from viewing the whole mechanism of the skeleton in constant relation to the final causes for which the different organs were contrived.'[23] Others, like Brougham, were equally impressed with the use of paleontological material in support of design in nature; though Thomas Turton, the regius professor of divinity at Cambridge, in a rather pedantic critique of Brougham, expressed 'some doubt whether the illustrations from *the remains* will render the reasoning from *design* more cogent'.[24]

Another Cambridge man who showed a certain degree of *jalousie de métier* was Sedgwick. He disapproved not so much of the substance but rather of the form of Buckland's natural theology. Sedgwick expressed his critical opinion in a long letter to Conybeare bringing to the surface the difference in style between the two geology professors of the English school. He wrote: 'The moral and theological part is, I think, a failure. In showing unity of design he is

[21] lvi (1836), 62.

[22] Ibid., liii (1835), 448.

[23] 17 Feb. 1837, *Proc. Geol. Soc.* ii (1838), 518. See also Lyell to Mantell, 14 June 1832, Mrs Lyell, *Life, Letters, and Journals of Lyell,* i (1881), 388. This letter does not refer to the BAAS meeting, but to one of the Geological Society on June 13 when a paper was discussed on the megatherium skeleton discovered by Woodbine Parish, *Proc. Geol. Soc.* i (1834), 403–4.

[24] *Natural Theology,* 1836, 54. On Brougham's book see Yule, 'Impact of Science on British Religious Thought', 1976, 187–213. Buckland aided Broughham to improve the *Discourse of Natural Theology.* See Buckland to Brougham, 26 Nov. 1838, 26 Mar. 1839, UC, Br P.

good; but the argument is broken up too much into fragments. He ought to have had one grand sweeping chapter on that head instead of 500 corollaries; several direct arguments fail of their aim. They ought to have appeared at the end, by way of removing objections.'[25] In essence Buckland loved the particular, whereas Sedgwick tended to generalize, as he had done in his *Discourse*.

Brewster, in spite of much Scottish hostility to Buckland's position of geological theory, agreed with the natural theology of the Bridgewater Treatise. In his essay for the *Edinburgh Review* he wrote:

Thus ennobled in its character, the natural theology of animal remains appeals forcibly to the mind, even when we consider these remains only as insulated structures dislodged from the interior of the earth; but when we view them in reference to the physical history of the globe, and consider them as the *individual beings* of that series of creations which the Almighty has successively extinguished, and successively renewed, they acquire an importance above that of all other objects of secular inquiry.[26]

This historical aspect of fossil occurrences provided natural theology with an altogether novel argument. Design in the world indicated a supreme designer; but this was only a refutation of atheism. Deists, who reduced the operations of nature to that of an autonomous machine, designed and set in motion only at the moment of its origin, could easily agree with the argument from design. For example, Nieuwentyt's *Regt Gebruik der Werelt Beschouwingen* had actually been popular with some of France's Enlightenment philosophers. However, their mechanistic worldview appeared to be confounded by historical geology with its evidence of not just one single beginning, but of a series of successive worlds, each the beginning of a new creation. Thus geology was able to refute not only the atheists but also the deists. Buckland, Conybeare, Sedgwick, and Whewell all emphasized this aspect of the natural theology of fossils. In his Trinity College sermon Sedgwick commented that earth history 'shews intelligent power not only contriving means adapted to an end: but at many successive times contriving a change of mechanism adapted to a change of external conditions; and thus affords a proof, peculiarly its own, that the great first cause continues a provident and active intelligence'.[27] Conybeare quoted Sedgwick approvingly,[28] and Whewell remarked 'that Geology has thus lighted a new lamp along the path of Natural Theology'.[29]

Buckland's *Geology and Mineralogy* represented a pinnacle of the design argument. Buckland's successors at Oxford, initially Hugh Strickland of dodo fame, and subsequently John Phillips, both employed natural theology retrospectively, appealing to the works by Buckland and others of the 1830s. In his public lecture 'On geology, in Relation to the Studies of the University of Oxford' (1850) Strickland remarked: 'Dr. Buckland's Bridgewater Treatise has completed the task which Paley had begun, and a quibbling mysticism is now the only resource of the sceptic and the atheist.'[30] Phillips, in his contribution to the *Replies to 'Essays and Reviews'* (1862) gave assurances that he recognized divine design in the geological record, 'in agreement with the conclusions of Conybeare, and the lectures of Buckland and Sedgwick'.[31]

### THE ORIGIN OF DEATH AND SIN

Although many of the new discoveries of geology added to the argument from design, they also formed the source of a new and emotive problem. The very existence of fossils indicated that death had occurred long before the appearance of man on earth. Furthermore, death had not only taken place by natural means, but violently, inflicted by one animal species on another. This was apparent from the carnivorous anatomy of certain vertebrate fossils, and it was most sensationally indicated by the composition of coprolites. Animal aggression in the geological past had been depicted with savage realism in the various reconstructions of ancient landscapes. The problem posed by this discovery of carnivorousness and death in the geological past derived from the traditional belief that such phenomena had not existed in the Garden of Eden and had entered the world because of, and subsequent to, the fall of man. This belief had found expression in, for example, Milton's *Paradise Lost*:

[27] *Discourse*, 1833, 23. See id. *Proc. Geol. Soc.* i (1834), 315–16.
[28] *Inaugural Address*, 1831, 34–5.
[29] *British Critic* ix (1831), 194. See also *Geology and Mineralogy*, i. 586.
[30] Jardine, *Memoirs of Strickland*, 1858, 212.
[31] 514.

Of man's first disobedience, and the fruit
Of that forbidden tree, whose mortal taste
Brought death into the world, and all our woe,
With loss of Eden . . .

                . . . Discord first
Daughter of Sin, among the irrational,
Death introduced through fierce antipathy:
Beast now with beast gan war, and fowl with fowl,
And fish with fish; to graze the herb all leaving,
Devoured each other;[32]

Old and New Testament texts were the basis for this belief that
suffering and death had entered the world as a consequence of
man's sin. In particular a statement by St. Paul in his letter to the
Romans seemed to indicate this: 'Wherefore, as by one man sin
entered into the world, and death by sin . . .'.[33] Biblical literalists,
opposed to the new perspective of pre-human earth history, were
quick to point out the discrepancy between geology and the Bible as
regards the origin of suffering and death. As early as 1817 a letter
was published in the *Philosophical Magazine* attacking Cuvier and
other geologists:

All orthodox Christians are agreed in this point: That if there had been no
sin, there would have been no suffering; that suffering of every kind is the
effect of sin; that Adam was constituted the head and representative of the
whole creation; and consequently that all the animals participated in the
consequences of his disobedience. But in this respect the Christian doctrine
is overturned, and, I may say, annihilated, by the system of geologists.
According to them, whole races of carnivorous animals inhabited both the
sea and the dry land before the creation of man; consequently the brute
creation must have been in a state of pain and suffering before Adam
fell.[34]

Both Buckland's *Reliquiae Diluvianae* and his Bridgewater
Treatise were followed by similar criticism from traditionalists.
Bugg, one of the most outspoken adversaries of Buckland and
Cuvier, asserted that '*Animals were not created carnivorous*'. He
continued: 'If animals were *created carnivorous*, "death," even
violent death must have been common in the creation from the very

[32] i. 1–4; x. 707–12.
[33] 5:12.
[34] Boyd, 'On Cosmogony', *Phil. Mag. and Journ.* 1 (1817), 377.

beginning. But the Scripture represents *death* as entering into the world by *sin*.'[35] George Young, J. Mellor Brown, and others objected to the new geology for the same reason.

In his Bridgewater Treatise Buckland devoted a separate chapter to the question of carnivorousness and suffering in nature; its title summarizes his argument: 'Aggregate of animal enjoyment increased, and that of pain diminished by the existence of carnivorous races.' Buckland reasoned from a utilitarian point of view, namely that God's goodness shows in the greatest amount of enjoyment for the largest number of individuals. This was achieved by the existence of carnivorous animals which functioned as a 'police of nature' eliminating the sick and the old who would otherwise have suffered as a result of pain and a lingering death. Carnivorous animals had also been a check on excess of numbers which would have produced a shortage of food and the starvation of many herbivores. Buckland grew lyrical in describing how carnage in the animal world had produced 'endless successions of life and happiness':

The appointment of death by the agency of carnivora, as the ordinary termination of animal existence, appears therefore in its main results to be a dispensation of benevolence; it deducts much from the aggregate amount of the pain of universal death; it abridges, and almost annihilates, throughout the brute creation, the misery of disease, and accidental injuries, and lingering decay; and imposes such salutary restraint upon excessive increase of numbers, that the supply of food maintains perpetually a due ratio to the demand. The result is, that the surface of the land and depths of the waters are ever crowded with myriads of animated beings, the pleasures of whose life are co-extensive with its duration; and which, throughout the little day of existence that is allotted to them, fulfil with joy the functions for which they were created. Life to each individual is a scene of continued feasting, in a region of plenty; and when unexpected death arrests its course, it repays with small interest the large debt, which it has contracted to the common fund of animal nutrition, from whence the materials of its body have been derived.[36]

This argument was by no means new: eighteenth-century natural theology had used it extensively to prove God's goodness,

<hr />

[35] *Scriptural Geology*, i (1826), 145.

[36] *Geology and Mineralogy*, i. 133–4. On the subject of evil in nature see also J. S. Duncan, *Analogies of Organized Beings*, 1831, 147; P. B. Duncan, 'Balance of Destruction and Preservation of Animals', *Literary Conglomerate*, 1839, 579 and *passim*.

in addition to his power and wisdom. The notion of a 'police of nature' to keep a balance of numbers occurred in Linnaeus's *Amoenitates Academicae*, argued by C. D. Wilcke (1760).[37] Paley had considered how God's goodness can be reconciled with the existence of venomous serpents and preying animals. He popularized the utilitarian argument that carnivorousness was no symptom of evil in nature but rather an indication of divine benevolence because it added to the sum total of animal enjoyment. Without it there would be a 'world filled with drooping, superannuated, half-starved, helpless and unhelped animals'.[38] Other writers on natural theology, especially in Scotland, followed this utilitarian line of reasoning, reminiscent of economics. Examples are Henry Fergus and John MacCulloch. The former wrote: 'The depredations of animals, then, upon each other is only an apparent, not a real evil: it forms no solid objection against the goodness of Deity, for it does not appear to diminish the happiness of any animal during life. It abridges suffering at death, and furnishes subsistence to a greater number of animals than could otherwise live on earth.'[39]

Even though Buckland's argument was not new, its application to the geological record and its enthusiastic presentation in the Bridgewater Treatise drew attention to it. But to some people this facile utilitarianism indicated a benevolence of character in Buckland rather than a divine goodness in nature. His ability to extract good out of evil, even out of the very fall of man, elicited satirical verse; 'The Apple' joked:

> If the Devil had failed, when Eve he assailed,
>   So sly, by the lure of an apple,
> We ne'er should have seen, in town, village, or green,
>   Cathedral, church, meeting, or chapel.

> It is true Paradise was delicious and nice,
>   Yet, if those born on earth had ne'er died,
> 'Twould have been such a cram, like the berries in jam,
>   Pic-a-back men and women must ride.[40]

The utilitarian argument may have provided a solution to the problem of evil in nature but not to the origin of death. How can

[37] vi (1763), 17–39.
[38] *Natural Theology*, ch. 26.
[39] *The Testimony of Nature and Revelation*, 1833, 301.
[40] Daubeny, 'Common Place Book', ii. 23–5, MC, Da P.

death be a punishment for man's sin if it occurred early in pre-human geological history? Buckland discussed this problem in a famous and controversial sermon preached on 27 January 1839 in the cathedral of Christ Church and published under the title *An Inquiry whether the Sentence of Death pronounced at the Fall of Man Included the whole Animal Creation, or was Restricted to the Human Race.* He had assumed in his Bridgewater Treatise that animal death was part of a 'universal law of mortality' and not an evil, but rather beneficent even when effected by carnivorousness. Now all he had to prove was that death and suffering were a punishment only to man and not to the creation as a whole. This was not contradicted by St. Paul's statement to the Romans that death (of man) was caused by sin (of man), and seemed to follow from another statement by St. Paul, to the Corinthians: 'For since by man came death, by man came also the resurrection of the dead.'[41] As the resurrection did not apply to animals, so their deaths should not be seen as in any way related to man's transgression.

Buckland further argued that no foundation can be found in the Bible for the belief that carnivorous animals did not exist in Paradise or ever will exist in a future world of peace and perfection. He concluded that 'we are free to conclude that throughout the brute creation death is in no way connected with the moral misconduct of the human race, and that whether Adam had, or had not, ever transgressed, a termination by death is, and always has been, the condition on which life was given to every individual among the countless myriads of beings inferior to ourselves, which God has been pleased to call into existence'.[42]

This attempt to solve the problem of death and human sin posed by the pre-human discoveries of geology made matters worse by bringing the problem out into the open. *The Times* published a number of letters on the subject of Buckland's sermon.[43] Margaret Ruskin, whose son John was then at Christ Church, wrote to her husband: 'John must explain to you how Dr. Buckland got over his difficulties without impugning the Scripture account. He, Dr. B. proves at least to his own satisfaction that the antedated animals (if one may so term them) eat, and were eaten by each other so that if they did not die naturally, they were killed. The question is a very

[41] 15: 21.
[42] 12.
[43] 9, 12, 13 Apr. 1839.

puzzling one—it is a great mercy that neither our welfare here nor our happiness hereafter depend on solving it . . .'.[44]

Buckland received sympathy from the Continent for his stance. The German chemist Bunsen, on a visit to England, wrote to his wife: 'Buckland is persecuted by bigots for having asserted that among the fossils there may be a pre-Adamite species. ''How!'' say they; ''is that not direct, open infidelity? Did not death come into the world by Adam's sin?'' I suppose then that the lions known to Adam were originally destined to roar throughout eternity!'[45] A similarly casual attitude toward the biblical account was shown by Deluc the younger, who reacted with interest to Buckland's sermon and added that he himself believed that St. Paul was not always to be relied upon.[46]

At home Buckland found an ally in the dissenter John Pye Smith. In the *Congregational Magazine* for 1837 Smith had already discussed the text on death and sin by St. Paul. Smith had concluded:

that it refers to the access and dominion of death *over man*, involving, the presupposition that, had not our first parents sinned, they would, on the expiration of their probationary state, have undergone a *physical change different from dying,* which would have translated them into a higher condition of happy existence. This glorious prospect they forfeited, and, as the just penalty of their transgression, sunk down into the condition of the inferior animals, in becoming the prey of temporal or corporeal death: but in relation to their higher capacities, they plunged themselves into the gulf of death in senses infinitely more awful.'[47]

Smith returned to the problem in *The Relation between the Holy Scripture and some Parts of Geological Science* (1839), and during the 1840s a number of other authors such as William McCombie and, across the Atlantic, Edward Hitchcock addressed the same question engendered by the 'astonishing disclosures' of geology.[48]

The issues raised by natural theology were never far from moral philosophy. Thus Paley, in addition to his *Natural Theology,* had written a major treatise on the *Principles of Moral and Political Philosophy* (1785). At Cambridge the same proximity of the two subjects existed, even though the utilitarian basis of Paley's moral

[44] 2 Feb. 1839, *Ruskin Family Letters*, ed. Bird, ii (1973), 583–4.

[45] Gordon, *Life and Correspondence of Buckland,* 1894, 136.

[46] Deluc to Buckland, 8 Aug. 1839, DRO, Bu P.

[47] *Relation between the Holy Scriptures and some Parts of Geological Science,* 1852, 325–6.

[48] E.g. McCombie, *Moral Agency,* 1842, 117.

philosophy was criticized. Sedgwick's *Discourse* discussed both the design argument in geology and Paley's moral philosophy. In 1845 Whewell published both his *Indications of the Creator* and a major work on the *Elements of Morality*. On a lower level P. B. Duncan added to an essay 'On the balance of destruction and preservation of animals' a talk on *Motives of War*. An author in the *Magazine of Natural History* argued that the moral quality of a government could be deduced from man's effect on the zoological balance in a country: a democratic government (not well regarded by the author) was likely to let any animal species, regardless of its usefulness or interest, be freely and wantonly persecuted by the human inhabitants; the monarchical government (much to be preferred) would act to regulate the animal population, contain pests, and propagate beneficial species.[49]

---

[49] Weissenborn, 'On the Influence of Man in Modifying the Zoological Features of the Globe', *Mag. Nat. Hist.* ii (1838), 13–18.

## 19
# The Idea of Progress

The geology of the English school was connected with such issues of wide interest as biblical exegesis and university reform. The natural theology of fossils put Buckland and his colleagues in yet closer contact with major issues of the day, particularly with the philosophy of happiness and pleasure and its ramifications in economic and social theory. The attempt to prove God's goodness, as well as his power and wisdom, on the basis of geological phenomena, turned the argument from design into an argument of moral philosophy. A system designed by God to a beneficial end gained *eo ipso* a divine seal of approval and was morally permissible if not imperative. Carnivorousness had contributed to the balance of nature and was thus morally justified; Buckland father and son made it the basis for a policy of zoophagy.[1] The notion of a police of nature was easily transferred to human society as a form of social control. Buckland used the gradation of animal life to justify the English class system in a sermon at Westminster in 1848 when political and social unrest threatened the status quo.[2]

Progressive development was the most comprehensive law of geology which the design argument turned into a justification of contemporary social belief. The progressivism of the English school was formulated at a time when the idea of progress was becoming a major determinant of cultural expectation in English society.[3] The geological notion of progressive earth history cannot be separated from this historical milieu. Loudon exclaimed in an editorial in his *Magazine of Natural History* (1834):

[1] Buckland is rumoured to have eaten, not only mice for breakfast, and ostrich for dinner, but the heart of a French king; see Morris, *Oxford Book of Oxford*, 1978, 193. See also Barber, *Heyday of Natural History*, 1980, 139–51.

[2] 'Equality of mind or body, or of wordly condition, is as inconsistent with the order of Nature as with the moral laws of God', *Sermon Preached in Westminster Abbey*, 1848, 13.

[3] See Bury, *Idea of Progress*, 1920; Pollard, *Idea of Progress*, 1968; Nisbet, *History of the Idea of Progress*, 1980.

So congenial are natural history pursuits to the human mind, and so much do they tend to the progress of civilisation, to increase domestic comfort, to peace between nations, and to human happiness, that to us it appears that it would be treason to nature to assert that this state of things will not be progressive, and will not go on increasing, till the condition of mankind every-where is improved to an extent of which we can at present form no idea.[4]

More can be demonstrated, however, than a general synchronicity of geological progressivism with a belief in social progress. In England progress was widely interpreted as an increase in material comfort. This definition was encouraged by the philosophical utilitarians and contrasted with that of German idealism, which saw progress as leading to moral perfection. The utilitarian interpretation made progressive development in the first instance a question of political economy. Geology and political economy interacted in the elaboration of the early nineteenth-century belief in progress, both at a cognitive level and, at Oxford, an institutional level.

As in nature, so in human society: a balanced system of economic and social forces was interpreted as divinely designed and thus should not be interfered with by capricious human government. An example of an apparently well-adapted system in society had been outlined by Thomas Malthus in his *Essay on the Principle of Population* (1798). He had argued that the number of people increases more rapidly than the means of subsistence, and that a balance is thus enforced by the permanently deprived state of the lower classes of society; i.e. poverty would adjust population to subsistence.

This Malthusian rule is well known if only because Darwin used it to formulate his mechanism of natural selection.[5] Long before Darwin, however, natural theology incorporated Malthusianism in its socio-economic theory. Sumner discussed 'the consistency of the principle of population with the wisdom and goodness of the Deity' in his *Treatise on the Records of Creation*. Malthusianism was championed by Thomas Chalmers, professor of divinity at Edinburgh, in his *Political Economy in Connexion with the Moral State and Moral Prospects of Society* (1832). He reiterated his views the following year in his more widely read Bridgewater Treatise. He argued that Malthusian population dynamics represented a

---

[4] vii (1834), iv.
[5] See Ospovat, *Development of Darwin's Theory*, 1981, *passim.*

balanced system and that the poor should not be artificially helped. Chalmers came down on the side of the landed proprietors, defending their divine right of property, on which labourers were little more than poachers. With the metallic intonation of Scottish Presbyterianism he commented:

> However obnoxious the modern doctrine of population, as expounded by Mr Malthus, may have been, and still is, to weak and limited sentimentalists, it is the truth which of all others sheds the greatest brightness over the earthly prospects of humanity—and this in spite of the hideous, the yet sustained outcry which has risen against it. This is a pure case of adaptation, between the external nature of the world in which we live, and the moral nature of man, its chief occupier. There is a demonstrable inadequacy in all the material resources which the globe can furnish, for the increasing wants of a recklessly increasing species.[6]

This view, expressed at the beginning of a decade of major Whig reform, was challenged in spite of its use of the design argument. Scrope took issue with Chalmers and the Malthusian school in his *Principles of Political Economy* (1833).[7] So did John Ramsey McCulloch, previously professor of political economy at the University of London. In his *Principles of Political Economy* (1843), he argued that man's capacity for improvement by way of industry and inventions should ultimately lead to the elevation of the lower classes from their state of poverty and deprivation.

Geology and political economy were concerned with the idea of progress to a greater extent than other subjects of academic study. Both had been recently added to English university curricula. The notion of progress in geology was analogous to that in political economy, and complemented it. Geological progress was defined by the English school as an increasing habitability of the surface of the earth to human society. In political economy progress was first and foremost defined as an increase in national wealth and consequently in public and private utility.

Geology and political economy came even closer in their environmental definition of progress where both were interested in the exploitation of mineral resources. The progressive habitability of the earth had been accompanied by the deposition of coal, iron ore, etc.; nature's progressive trend in building up the crust of the earth

---

[6] *Adaptation of External Nature to the Moral and Intellectual Constitution of Man*, 1833, ii. 49.

[7] See Rudwick, 'Poulett Scrope on the Volcanos of Auvergne: Lyellian Time and Political Economy', *Brit. Journ. Hist. Sci.* vii (1974), 205–42.

was now to be continued by man's own effort in the exploitation of these natural resources.

This connection of geology and political economy in the definition of progress was made by Tennyson in his *In Memoriam* (1850).[8] Progress had already come to mean an increase in national prosperity, with Adam Smith's classic *Inquiry into the Nature and Causes of the Wealth of Nations* (1776). This definition was echoed by the voluminous body of literature on political economy which accumulated during the early part of the nineteenth century. McCulloch's catalogue of the *Literature of Political Economy* (1845), which exceeded 400 pages in length, defined progress as a change 'From poverty and barbarism to wealth and civilisation'.[9] But this progress was based on a particular system of capitalist economy. In England political economy in tandem with geology used the design argument to justify not only the drive for progress but also the economy of free enterprise and *laissez-faire*.

Good examples of this are to be found in the works of Buckland and Whately, the latter the second holder of the Drummond chair of political economy at Oxford, from 1830 till 1832, in succession to the first holder Nassau William Senior (1825–30). Nares had already given some lectures on the 'Importance and Nature of the Science of Political Economy' (1817, 1820).[10] Buckland had been interested in political economy from the early days of his international travels, and Whately was one of the Oriel noetics who attended Buckland's geology lectures, and was the author of a satirical 'Elegy intended for Professor Buckland'. (He also wrote the following rebuke to Buckland for the latter's infamously illegible handwriting: 'Dear Buckland, On our return last night I found as I thought that a spider had crawled out of the inkstand over a bit of paper; but it turns out to be a hieroglyphic from you, which I so far interpreted as to perceive it was an invitation to meet some Professor, whose name as you wrote it looked somewhat indecent. I shall be happy to wait on you and take the opportunity of learning the Egyptian mode of writing.'[11])

Buckland and Whately were faced with the same problems in getting their new subjects accepted at Oxford and in drawing a

[8] *Poems of Tennyson,* ed. Ricks (1969), 969–70. Quoted above, p. 231.
[9] 77. See also Mathias, *First Industrial Nation,* 1969, 291.
[10] Bodleian Library, MSS Top. Oxon. d. 357–8.
[11] Undated, DRO, Bu P.

student audience from tutorial college rooms to the university lecture theatres. Whately specifically identified with geology in his *Introductory Lectures on Political Economy* (1831). He asserted that the Bible should not be used as a textbook either for geology or for political economy, but that this does not set these new subjects at variance with theology. On the contrary; Whately, sympathetic to Chalmers's Malthusianism, argued that political economy provides some of the finest and most remarkable examples of design to the theology student, like the balance of supply and demand in a free economy. Shortly before his appointment Whately wrote to a friend that he was thinking 'of making a sort of continuation of Paley's ''Natural Theology'', extending to the body-politic some such views as his respecting the natural'.[12]

To Whately the ultimate illustration of design was economic progress. 'In nothing, perhaps, will an attentive and candid inquirer perceive more of this divine wisdom than in the provisions made for the progress of society.'[13] He regarded a capacity for improvement as characteristic of the human species, both as individuals and as communities. Whately commented that the laws of physics and chemistry do not vary and that the instincts of animals do not lead to improvement of their state. 'But in man, not only the faculties are susceptible of much cultivation, (in which point he does indeed stand far above the brutes, but which yet is not *peculiar* to our species,) but besides this, what may be called the *instincts* of man lead to the advancement of society.'[14] Whately's protégé William Cooke Taylor, a regular contributor to the *Athenaeum*, developed the idea of human improvement in his *Natural History of Society* (1840).

To Buckland and the English school a divine purpose unfolded in the course of geological history directed towards man and his dominion of the earth. In this teleological perspective of earth history human progress was a temporal extension and continuation of geological progress, divinely ordained in its congruity with the past. However, human history was more than just a temporal continuation of geological history; it represented a recapitulation of the geological past. The history of man was not only the last period

---

[12] E. J. Whately, *Life and Correspondence of Whately*, i (1866), 67. See also Rashid, 'Richard Wately and Christian Political Economy', *Journ. Hist. Ideas.* xxxviii (1977), 147–55.

[13] *Introductory Lectures*, 101.

[14] Ibid., 112.

of a succession of periods of earth history, but it constituted also the culmination, even the consummation, of all previous worlds. This rather metaphysical interpretation of history was not spelled out in publications, but it was put on paper in correspondence and lecture notes. J. S. Duncan's 'Moral view of the earth's strata' interpreted the geological rock succession in terms of human history:

Granite in its well separated and crystallized substances typifies the Paradisaical state; Gneiss, primitive corruption; Slate and Graywacke, early and imperfect social order; Old Red Conglomerate the great wreck of nations; metallic strata, coals and mountain lime the struggles of civilization amidst wars and heart burnings and strifes; Lias and Oolite and Green Sand and Chalk the superior order of Civilization and the ruins of states from Solon to the sixteenth Century; the gravel beds and other strata above the Chalk (including peat) modern wars and modern extrication of art scrap by scrap from antique fragments.[15]

Buckland seriously believed in an analogy between the main periods of geological history and those of human history, the latter being a microcosm of the geological past. To Agassiz the embryonic growth of an animal recapitulated the geological development of its class. To Buckland human history recapitulated the main periods of earth history and represented its purpose. He believed that the four main geological periods, Primary, Transition, Secondary, and Tertiary, correspond to four epochs of man's history, namely mythological prehistory, ancient history, the Middle Ages, and modern times.[16] He could joke therefore about dinosaurs living during the Secondary and dragons like those of St. George during the Middle Ages.[17] He could also use part of the old Transition rocks as a metaphor of Conservative politics, writing to Murchison: 'I promised Lord Lyttelton that like a good Tory you would work in the strata beneath the coal.'[18] This human recapitulation of geological progress must have been intended when Tennyson wrote that man would be 'in higher place, if so he type within himself this work of time, from more to more'.

[15] Quoted by P. B. Duncan to Buckland, undated, OUM, Bu P.

[16] Stratigraphy file, OUM, Bu P. The idea was not new; it occurs in Blumenbach, *Beyträge zur Naturgeschichte,* 2nd edn 1806, i. 115–16. See also Rudwick, 'Historical Analogies in the Geological Work of Charles Lyell', *Janus,* lxiv (1977), 89–107.

[17] 'Dragons in middle ages of Geology, as well as in the romances of historical middle ages', Jackson, 'Buckland's Geological Lectures, 1832', 1 IGS, Bu P. In an autograph outline of lecture topics the following occurs: 'Analogy of Geological Chronology to the four historical periods', Stratigraphy file, OUM, Bu P.

[18] 19 June 1835, NMW, Bu P.

This view of human history as the fulfilment as well as the continuation of the geological past was implicit in Buckland's discussion of the adaptation of mineral resources to human needs. Political economy argued that man's improvement was divinely ordained and that it was based, above all, on industry and its exploitation of mineral wealth. Geology showed how the very disposition of coal, iron ore, etc., inside the crust of the earth indicated divine contrivance to facilitate human exploitation for the benefit of modern society.

Some people argued that the sole purpose of all geological history and of all nature was man's well-being. The German geochemist Bischof actually speculated that the purpose of life in the geological past had been to become fossilized and thus form a primeval manure for the later benefit of human agriculture.[19] Like Paley, Buckland argued that animal life had its own purpose, and he urged 'to consider each animal as having been created first for its own sake, to receive its portion of that enjoyment which the Universal Parent is pleased to impart to each creature that has life; and secondly, to bear its share in the maintenance of the general system of co-ordinate relations, whereby all families of living beings are reciprocally subservient to the use and benefit of one another.'[20]

However, a purpose more directly related to man could be argued for the inert minerals of the earth's crust. Buckland ended his Bridgewater Treatise with no fewer than five chapters illustrating how the composition and structure of the earth's crust had been purposefully designed for the benefit of mankind, particularly for nineteenth-century society and most generously for Great Britain in support of its industry and Empire. Two features in particular showed the beneficent design (figure 14); first, the formational juxtaposition of the three lithologies needed for iron smelting: coal, iron ore, and limestone; and secondly, the basin-shaped nature of the coal measures and of other geological successions.

The argument was not new: Buckland had referred to it in his *Vindiciae Geologicae*, and Conybeare had spelled it out in his contribution to the Carboniferous section in *Outlines of the Geology of England and Wales*. The occurrence of iron ore in immediate proximity to coal, the fuel needed for its reduction, and the limestone

---

[19] 'On the Terrestrial Arrangements Connected with the Appearance of Man on the Earth', *Edinburgh New Phil. Journ.* xxxvii (1844), 49.

[20] *Geology and Mineralogy*, i (1836), 101.

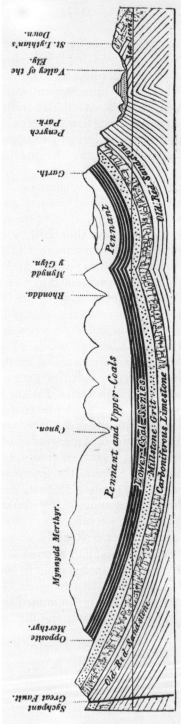

14. Vertical cross section from north to south through the coal basin of South Wales showing the geological juxtaposition of coal measures with ferruginous shale beds, millstone grit, suited for the construction of furnaces, and limestone, needed for the reduction of iron ore.

which facilitates that reduction 'is an instance of arrangement so happily suited to the purpose of human industry, that it can hardly be considered as recurring unnecessarily to final causes'.[21] Buckland formulated the argument more directly: the geological proximity of coal, iron ore, and limestone indicated a divine contrivance for the furtherance of industry, especially for the iron smelting works clustered in and around Birmingham, South Wales, Derbyshire, Yorkshire and the south of Scotland. Buckland extolled the uses of coal and iron, 'those two fundamental elements of art and industry, which contribute more than any other mineral production of the earth, to increase the riches, and multiply the comforts, and ameliorate the condition of mankind.'[22]

And, however remote may have been the periods, at which these materials of future beneficial dispensations were laid up in store, we may fairly assume, that besides the immediate purposes effected at, or before the time of their deposition in the strata of the Earth, an ulterior prospective view to the future uses of Man, formed part of the design, with which they were, ages ago, disposed in a manner so admirably adapted to the benefit of the Human Race.[23]

Buckland argued that the basin-shaped occurrence of the coal measures was a further feature of beneficent design. For it facilitates mining by bringing the coal beds to the surface around the circumference of each basin, making them accessible to human mining technology; an uninterrupted dip in one direction only would have soon plunged the strata to a depth inaccessible to man. Buckland followed a similar line of reasoning for the London, Paris, and Vienna basins, where the curved disposition of the Chalk and the Tertiary beds make possible the sinking of artesian wells for the water supply of the populous cities. Even such apparently random features as faults in rocks could be understood as beneficently designed: for example, where a dislocation had off-set a particular coal bed, an accidental fire would be prevented from spreading across the fault.

This application of natural theology reinforced the early Victorian self-confidence which was encouraged to believe that national prosperity and power were divinely ordained and that to work for their increase was a moral imperative. The cultural

[21] 333; see 234.
[22] *Geology and Mineralogy*, i. 67.
[23] Ibid., 536–8.

exuberance fostered by geology and political economy was apparent in McCulloch's *Statistical Account of the British Empire* (1838) which described its 'extent, physical capacities, population, industry, and religious and educational institutions'; Robert Bakewell contributed the geology sections.

Buckland's use of the argument from design expressed a dominant trend of socio-economic opinion; the early Victorians revelled in his geological proofs of God's generosity towards the British Isles. Scrope, for one, himself a geologist and author of the anti-Malthusian though utilitarian *Principles of Political Economy*, approvingly and extensively quoted from the last few chapters of Buckland's Bridgewater Treatise in his review for the *Quarterly Review*.

The *Mining Review* reacted enthusiastically to Buckland's application of the design argument to mineral deposits. It praised the Bridgewater Treatise in a long essay as the most prominent work of the year connected with mining activity.[24]

The popularity of the mineral design argument was described by a German naturalist who visited the Birmingham meeting of the British Association in 1839. The nearby limestone caverns of Dudley, artificially excavated for the iron smelting works, were illuminated for the occasion, and a visit to the caverns was arranged. The main vault, about a mile long, one hundred feet high, and seventy-five feet wide, was filled with thousands of visitors from the surrounding region. First Murchison, with a stentorian voice reminiscent of his military past, addressed the crowd on the geological condition of the area. After him 'Buckland went to the gallery, placed himself on a mighty block of stone, and lectured for more than an hour, he and his numerous audience being veiled in the wreathing sulphur smoke, upon the subject already handled by Murchison, but in so original and humorous a manner that he held the attention of his listeners in a way seldom witnessed.'[25] The upshot of Buckland's speech was:.

The immeasurable beds of iron-ore, coal, and limestone which are found in the neighbourhood of Birmingham, lying beside or above one another, and to which man has only to help himself in order to procure for his use the most useful of all metals in a liberal measure, may not, he urged, be

[24] 1837, 72–104.
[25] Gordon, *Life and Correspondence of Buckland*, 1894, 81.

considered as mere accident. On the contrary, it in fact expresses the most clear design of Providence to make the inhabitants of the British Isles, by means of this gift, the most powerful and the richest nation on earth.[26]

The general popularity of Buckland's utilitarian application of the design argument was not matched by an equal approval among his academic colleagues. Agassiz commented sardonically that it seemed risky to assume that coal measures had originated in the geological past for the special benefit of English industry today.[27]

Buckland never addressed the issue of utilitarianism formally, but his justification for carnivorousness and death was purely Benthamite. He specifically used the language of political economy to argue that when a creature is suddenly and unexpectedly killed by an animal of prey 'it repays with small interest the large debt, which it has contracted to the common fund of animal nutrition, from whence the materials of its body have been derived'.[28]

Sedgwick and Whewell disapproved of Bentham's utility or Paley's expediency as a basis for moral philosophy. Sedgwick abhorred the consequences of utilitarianism when transposed on to the human plane where it reduced the morality of an act to its material consequences. 'From first to last,' he pronounced, 'it is in bondage to the world, measuring every act by a wordly standard, and estimating its value by wordly consequences. Virtue becomes a question of calculation—a matter of profit and loss; and if man gain heaven at all on such a system, it must be by arithmetical details.'[29] To Sedgwick and Whewell man has an innate sense of right and wrong. Whately also argued in his Dublin sermons that we have a moral faculty implanted by God.[30] Bulwer-Lytton, educated at Trinity College, denounced what he called 'the economic philosophy'. He commented sarcastically that in the Middle Ages the burning of witches provided the greatest happiness to the greatest number: '*their happiness* demanded a bonfire of old women'.[31]

This difference about utilitarian morality extended to economic geology and the practical application of geological knowledge.

[26] Ibid., 82.

[27] *Geologie und Mineralogie,* i (1839), 602–3.

[28] *Geology and Mineralogy,* i. 134.

[29] *Discourse,* 1833, 57. See Garland, *Cambridge before Darwin,* 1980, chapter 4.

[30] See McKerrow, 'Richard Whately on the Nature of Human Knowledge', *Journ. Hist. Ideas.* xlii (1981), 439–55.

[31] *England and the English,* ii (1833), 111.

Neither Sedgwick nor Whewell shared Buckland's enthusiasm for the public utility of geology. This is well illustrated by the different interpretations of the term 'mineralogy'. To Buckland it meant the study of mineral resources; to Sedgwick and Whewell it applied to the more abstruse chemistry and crystallography of mineral substances. This is how Whewell interpreted the subject in his report to the British Association 'On the recent progress and present state of Mineralogy' (1832). Sedgwick wrote about Buckland's Bridgewater Treatise to Conybeare that 'the chapter on Mineralogy is decidedly bad and jejune. Had he read Whewell's philosophical *Report* he must have seen where Mineralogy was among the sciences, and what argument it might supply.'[32]

[32] 5 Dec. 1836, Clark and Hughes, *Life and Letters of Sedgwick*, i. (1890), 470.

# 20
# Oxford's Tractarians against Geology

Geology had justified its existence at Oxford by showing that it was a branch of natural theology. Ironically this form of self-justification contributed to geology's decline at the university during the 1830s and 1840s. The meeting of the British Association in 1832 which Buckland concluded by giving his talk on the megatherium represented a peak of geological success at Oxford. The same meeting aggravated traditionalist opposition to science, which coalesced to form the Oxford Movement, led by men like Froude, Keble, Newman, and Pusey. They regarded the argument from design as wholly inadequate and disliked its latitudinarian implications. The anti-scientific attitude of the Tractarians cast a cold shadow over the still young subject of geology, retarding its further growth at the university.

Tractarian opposition represented a volte-face. The leading members of the Oxford Movement had been elected to Oriel College fellowships during Copleston's time as provost (1814–28) together with such noetic latitudinarians as Arnold, Hampden, and Whately. Copleston's young fellows had been favourably disposed to the introduction of new subjects like geology and political economy. Most of them, even Keble, had attended Buckland's lectures. Some had joined the Ashmolean Society in 1828.[1] Pusey contributed to Buckland's Bridgewater Treatise. Isaac Williams, who became a close friend of both Keble and Newman, gained his early reputation as a poet by winning the chancellor's prize for Latin verse on the subject of geology.

The Oxford Movement originated as a reaction to the Whig reform of the 1830s, already preceded by such examples of Tory reform as the repeal of the Test Acts (1828) and the Catholic

---

[1] See the register of attendance at Buckland's lectures, OUM, Bu P. Also 'Lists of Members of the Ashmolean Society', ibid. Newman's name occurs as late as 1835, but in 1836, the year of the Hampden affair, his membership was terminated.

Emancipation Act (1829).[2] Peel failed to gain re-election as a member of Parliament for Oxford because of his reformist stance. The issue divided such former Oriel friends as Newman and Whately.[3] The passage of the reform bills of 1832, compounded by the interdenominational meeting of the British Association at Oxford, seemed to bring closer a possible repeal of the religious texts at the university, thus opening its gates to dissenters and undermining its character as the nursery of the established Church. Keble expressed his concern in his Assize Sermon of 14 July 1833. Similar traditionalist concern was expressed in a numerous series of *Tracts for the Times* begun by Newman. The *British Critic* became the national platform for the Tractarian critique of science, its institutions, and its use of the design argument.

The Oxford Movement was essentially High Church in character, emphasizing the catholicity of the English Church and the continuity of its apostolic succession. This interest in the unity and history of the church inevitably caused the Tractarians to drift in the direction of Rome. Accusations of crypto-Catholicism were common and seemed corroborated when in 1845 Newman actually turned Roman Catholic. Buckland disparagingly referred to his former friend Pusey as 'the monk'.[4] The Tractarians' ideal included a doctrinally sharply defined and narrow basis for Oxford and its education, with an emphasis on the authority of tradition and on spiritual values.[5]

This ideal clashed with that of science and with its self-justification by contributions to the design argument. Natural theology, it was believed, could prove the existence of God, and could also illustrate some of his attributes, but it was manifestly incapable of providing convincing proofs of Christianity, let alone of the doctrinal points which divided its many denominational expressions. Thus a variety of religious denominations could shelter under the umbrella of the design argument. This was exemplified by scientific periodicals and institutions. Loudon asserted in one of his editorials to the *Magazine of Natural History* that a taste for science interferes with no political or religious interest.[6] Such latitudin-

---

[2] On reform see Brock, *Great Reform Act*, 1973.

[3] See E. J. Whately, *Life and Correspondence of Whately*, i (1866), 64.

[4] Buckland to Featherstonhaugh, 14 Oct. 1845, CUL, Se P.

[5] See Faber, *Oxford Apostles*, 1936, *passim*; Church, *The Oxford Movement*, ed. Best, 1970, *passim*.

[6] vi (1833), 'Preface'.

arianism characterized institutions like the Geological Society of London, and it was the basis on which William Harcourt organized the British Association. As George Peacock, dean of Ely, commented in his presidential address to the Association:

> The founders of the British Association justly conceived that men of different shades of political opinion or religious belief would rejoice in the opportunities which such Meetings would afford them of coming together, as it were, upon neutral ground, where their mutual warfare would, for a season at least, be suspended, and no sounds be heard but those of peace: they felt persuaded that the softening influence of reciprocal intercourse would tend to soothe the bitterness of party strife, and would expose to view points of contact and union even between those whom circumstances had most violently estranged from each other, and show them that the features of the monsters of their apprehension were not so repulsive as their imaginations or intolerance had drawn them.[7]

This interdenominational tolerance was brought into Oxford by Buckland when he presided over the first full meeting of the British Association in 1832. The conservative *Oxford University Magazine* in 1834 attributed the early success of the association to Buckland, 'who with a generosity almost chivalrous, invited the infant body to the hospitable halls of Oxford. Here its numbers doubled, and the celebrity of the place gave celebrity to the institution. The same result occurred at Cambridge.'[8] The latitudinarian character of the event was emphasized by the conferment of honorary degrees on four nonconformists, Brewster, Brown, Dalton, and Faraday, a move welcomed by the liberal press which saw it as a step towards opening the ancient universities to full participation by dissenters.[9]

This is also how the Tractarians regarded the event, though with shock and horror. Keble denounced Buckland's speeches as 'plain idolatry'. The *British Critic* recognized and deplored the latitudinarian character of the British Association, its meeting in Oxford, and the interdenominational generality of the design argument. Newman reminded his readers that Oxford was a creation of the Middle Ages, that Oxonians were 'the Benedictines and Augustinians of a former day', and that the university should not give in to modern fashion, as represented by the British Association. He feared that its

[7] *Report BAAS, 1844*, xxxiii.
[8] i (1834), 402.
[9] *Fraser's Mag.*, v (1832), 751.

meeting in Oxford had brought the university closer to the admission of dissenters:

Scarecely had a twelve month passed, when the proper fruits of it appeared; those who had been admitted to covet, felt a greater pang at its gates being closed against them, than pleasure in the memory of the short week when they had been opened; and the visit of the savans to Oxford was the precursor of the bill introduced into the commons for the permanent admission of Dissenters to its lecturerooms. Such is the inevitable consequence of aping or trembling at the external world.[10]

The memory of the traumatic incursion into Oxford of scientific latitudinarianism was kept alive among the Tractarians long after 1832. Thomas Mozley felt compelled as late as 1882 to vilify Buckland for his part in the British Association meeting. Some unfavourable references to Buckland in Mozley's *Reminiscences, chiefly of Oriel College and the Oxford Movement* triggered a controversy, and Mozley explained himself the same year in a *Letter to the Rev. Canon Bull*; he summed up why the Tractarians had resented Buckland so much: 'Buckland was at Oxford the representative of science, and the *savans* of the University sheltered themselves under his great name. They certainly considered themselves at war, I will not say with faith, but with a large mass of secondary beliefs, and, what is more, with a number of pious traditions, cherished the more fondly because appealing rather to loyalty than to faith.'[11]

Buckland's identifications of a saint's skeleton as that of a goat and of a saint's blood as bat's urine must indeed have hurt the pious feelings of the Tractarians who valued tradition more than scientific fact. To Mozley Buckland's eighteenth-century sense of humour was not only coarse but profane. Recalling the meeting of the British Association of 1832, he wrote: 'At the *soiree* of the British Association at Oxford, I saw Buckland, of course a prominent personage in the room, and surrounded by an admiring circle, take out of his pocket a white handkerchief with a life-sized portrait of Queen Adelaide on it, display it, blow his nose with it, and replace it in his pocket, eliciting much laughter at every stage of the exhibition.'[12] Mozley believed that a man who could do this to royalty could not be far away from disrespect for divinity and thus from profanity.

[10] xxiv (1838), 145. See also Morrell and Thackray, *Gentlemen of Science,* 1981, 224–36.
[11] 20–1.
[12] Ibid., 22.

Mozley also commented on Buckland's Bridgewater Treatise. He praised its illustrations of design, but 'We all know that much more is requisite.'[13] The *British Critic* stressed that the soundest foundation of religious faith is the authority of tradition, represented by the clergy or by one's elders. The evidence of natural theology adduced by Paley and by the Bridgewater Treatises was denounced as inferior, and the reviewer made the following assessment, which was confirmed later by the triumph of Darwin's theory of natural selection: 'The whole fabric of Catholic Christianity has been shifted bodily, and without awakening its inhabitants, from the sound old piles of authority on which its Founder placed it, to new fantastic props rotten in themselves, and half sawn through by those who framed them, and who are now waiting to see the crash.'[14]

Less emotive criticism of natural theology was to be found in Newman's *Idea of a University* (1852–8). The design argument taught God's power, but 'What does Physical Theology tell us of duty and conscience? of a particular providence? and, coming at length to Christianity, what does it teach us even of the four last things, death, judgment, heaven, and hell, the mere elements of Christianity? It cannot tell us anything of Christianity at all.'[15] Thus the contributions of science to natural theology were no ground for its inclusion in university teaching; nor were its economic applications. Newman did not deny the attraction as well as the benefit of subjects such as geology, but he asserted that science does not have the same merit as classics in providing a 'robust and invigorating discipline for the unformed mind'.[16]

This anti-scientific attitude suited the classicists, who had long felt that geology encroached on their territory. Thomas Gaisford, dean of Christ Church and a classical scholar of considerable repute, thanked God on Buckland's departure for a continental journey that they would hear no more of his geology: let the young men concentrate on Greek literature which not only elevates above the vulgar herd but may lead to positions which carried considerable emoluments. Benjamin Jowett, master of Balliol and regius

[13] Ibid., 24.
[14] (Sewell) xxiv (1838), 306. See 'List of Contributors to the British Critic, 1836–41', PH. Ne P.
[15] Ed. Kerr, 1976, 365.
[16] Ibid., 222.

professor of Greek, later expressed a similar sentiment, seeing science as a menace to the 'higher conception of knowledge and of the mind'.[17]

In the eyes of the Tractarians science not only compromised itself in 1832 when it received the British Association, but it compounded its disloyalty to the Anglican integrity of Oxford in 1836 by siding with Renn Dickson Hampden. In a footnote to his *Essay on the Philosophical Evidence of Christianity* (1827) Hampden had expressed the Evangelical opinion that the Thirty-nine Articles, the doctrinal identity of the established Church, carried authority only to the extent that they expressed biblical truth: "The Articles of the Church of England not consisting so much of affirmations of scripture truth, as of negations of doctrines unscripturally introduced into the body of faith; it is evident, that their whole drift is, to maintain the *exclusive* authority of scripture, and not to limit it by selection.'[18] This view of the Thirty-nine Articles appeared to encourage interdenominational tolerance; and was the fruit of the noetic school of Copleston's Oriel, warmly supported by Arnold and Whately. Hampden did nothing to detract from his latitudinarian position in his Bampton lectures of 1834. Two years later the Whig government of Melbourne, on Whately's recommendation, appointed Hampden to the regius chair of divinity. A storm of protest broke loose over Oxford. Tractarians such as Mozley, Newman, and Pusey accused Hampden of doctrinal impurity. When his appointment could not be blocked, they tried to limit his powers as divinity professor by keeping him off the board of inquiry into heretical affairs and the board for the nomination of select preachers.[19]

P. B. Duncan wrote a circular in support of Hampden warning against 'uncharitable bitterness of controversy'.[20] Buckland tried to keep out of the conflict, but to no avail. Bakewell wrote to Silliman: 'Oxford, where B. resides, and is a Canon, has been thrown into a great ferment about Dr. Hampden's free opinions, and the geologists have come in for a share of the censure.'[21]

---

[17] Abbott and Campbell, *Life and Letters of Jowett*, ii (1897), 268. See Tuckwell, *Reminiscences of Oxford*, 1907, 55.

[18] 297–8.

[19] See n. 5.

[20] 'Address to Members of Convocation', 2 May 1836, 'Hampden Controversy', 1836, a collection of pamphlets in the Bodleian Library.

[21] 28 July 1836, Fisher, *Life of Silliman*, ii (1866), 56.

The Tractarians drew large numbers of students to their sermons; Whately pessimistically estimated that by October 1838 no fewer than two-thirds of the students were 'Puseyite'.[22] Newman in particular commanded a considerable following.

Newman divided Oxford. The old and the buttoned were not converted. But the young fell under a thrall. The undergraduates first went to Newman because he was disreputable among their elders, because his name was exciting, because he banged the regius professor, because the chaplain of New College placarded Oxford against his popery. They stayed to discover an ethical power which led them to examine the unwonted doctrine and then to revere the teacher. From the pulpit of St. Mary's they learned obedience, holiness, devotion, sacrament, fasting, mortification, in language of a beauty rarely heard in English oratory.[23]

The attendance at science lectures dropped as that at Newman's sermons soared. Buckland's rocks and fossils, his anecdotal idiosyncracy, and his Benthamite proclivity, were no longer a match for the early Victorian earnestness and spiritual conviction of Newman. Even the appearance of Buckland's Bridgewater Treatise caused no more than a brief upturn in student attendance at the geology lectures. Comparison of the 1820s with the 1830s shows that all science courses suffered; the numbers for the anatomy lectures decreased from twenty-nine to seventeen and a half *per annum*; for chemistry from thirty-one to sixteen; for physics from forty-two to about ten; and for geology, though best attended of all, the drop was still marked, from fifty to thirty.[24] The scientists reacted by asking for compulsory lecture attendance and for a separate honours school of natural science. Daubeny in particular fought to reach these goals.[25] In 1847 he, Henry Acland, and others started a movement for a university museum to bring together the various natural history collections. They invited Buckland to add his signature to a gravamen, but by this time Buckland had despaired of Oxford: he refused.

Some years ago I was sanguine, as you are now, as to the possibility of Natural History making some progress in Oxford, but I have long come to

[22] E. J. Whately, *Life and Correspondence of Whately*, i. 418.

[23] Chadwick, *Victorian Church*, i (1971), 168–9.

[24] See Daubeny, 'To the Members of Convocation', 24 Feb. 1839, 11 Mar. 1839, *Papers Relating to the Proceedings of the University*, 1839, Bodleian Library.

[25] E.g. *Brief Remarks on the Correlation of the Natural Sciences*, 1848; also the anonymously published *Dream of the New Museum*, 1855.

the conclusion that it is utterly hopeless. The idle part of the young men will do nothing, and the studious portion will throw their attention into the channel of honours and profits which can alone be gained by the staple subjects of examination for degrees and fellowships. At present it is a detriment to a candidate for either to have given any portion of his time and attention to objects so alien from what is thought to tbe the proper business of the university as natural history in any of its branches.[26]

Buckland left Oxford in 1845 to become dean of Westminster. In London he indulged in his long-suppressed interest in public utility, a cause he already championed in his presidential addresses to the Geological Society in 1840 and 1841.[27] Of his close colleagues only Daubeny at Oxford and de la Beche in London followed a similar utilitarian trend, the latter *ex officio*, as he had become the director of the Geological Survey (1835). Conybeare and Whewell were no longer much involved in geology; Murchison and Sedgwick became locked in stratigraphic quarrels of a specialized nature; and Owen took Buckland's mantle as England's foremost vertebrate paleontologist. When Buckland, in 1840 and 1841, attempted to rally his old allies behind Agassiz's glacial theory, he encountered scepticism and opposition. A new ally was Thomas Sopwith, a civil engineer and mining surveyor from Newcastle.

During the 1840s Buckland contributed to such areas of economic geology as agriculture, sanitation and water supply, architecture, and civil engineering.[28] In recognition of his patronage, he was made honorary member of the Agricultural Society and the Institution of Civil Engineers. *Punch* depicted Buckland as 'Professor Buckwheat educing the agricultural mind' on improvements of drainage, fertilisers etc.[29] The final phase of Buckland's career represented an abandonment of academe and of the English school and reflected a utilitarian drift for which his Bridgewater Treatise had outlined the divine justification and which culminated in the Great Exhibition of 1851, not long before Buckland's death in 1856. Thus his last active decade put into practice Tennyson's couplet from *Locksley Hall* (1842): 'There methinks would be enjoyment more than in this march of mind, In the steamship, in the railway, in the thoughts that shake mankind.'[30]

[26] Buckland to Acland, 27 Dec. 1847, RSL, Bu P.

[27] 21 Feb. 1840, 19 Feb. 1841, *Proc. Geol. Soc.* iii (1842), 211–24, 471–6.

[28] See Buckland's review of 'Agriculture', *Quarterly Review*, lxxiii (1844), 477–509; his *Address to the Artesian Well Committee*, 1844; and his *Sermon on the Removal of the Cholera*, 1849.

[29] 20 Aug. 1845, 98.

[30] *Poems of Tennyson*, ed. Ricks (1969), 698.

# Bibliography

MANUSCRIPTS

| Collection | Location | Abbreviations |
|---|---|---|
| Brougham | University College, London | UC, Br P |
| Buckland | Bodleian Library, Oxford | BL, Bu P |
| | Christ Church, Oxford | CC, Bu P |
| | Devon Record Office, Exeter | DRO, Bu P |
| | Fitzwilliam Museum, Cambridge | FM, Bu P |
| | Institute of Geological Sciences, London | IGS, Bu P |
| | National Museum of Wales, Cardiff | NMW, Bu P |
| | Radcliffe Science Library, Oxford | RSL, Bu P |
| | Royal Society, London | RS, Bu P |
| | University Museum, Oxford | OUM, Bu P |
| Conybeare | National Museum of Wales, Cardiff | NMW, Bu P |
| Cuvier | Museum National d'Histoire Naturelle, Paris | Cu P |
| Daubeny | Magdalen College, Oxford | MC, Da P |
| de la Beche | National Museum of Wales, Cardiff | NMW, Bu P |
| Greenough | Cambridge University Library | CUL, Gr P |
| Henslow | Cambridge University Library | |
| Newman | Pusey House, Oxford | PH, Ne P |
| Peel | British Library, London | BL, Pe P |
| Phillips | University Museum, Oxford | |
| Sedgwick | Cambridge University Library | CUL, Se P |
| | Sedgwick Museum, Cambridge | SM, Se P |
| Smith | University Museum, Oxford | |
| Whewell | Trinity College, Cambridge | TC, Wh P |
| Winch | Linnean Society, London | LS, Wi P |
| Wollaston | Cambridge University Library | CUL, Wo P |

This list does not include relevant papers at the Geological Society, London, to which the author was refused access.

PRIMARY PRINTED SOURCES

The majority of anonymous and small news items and book reviews taken from the following periodical publications are not separately listed in the bibliography:

*American Journal of Science*
*Annals of Science*
*Arcana of Science*
*Athenaeum*
*Edinburgh (New) Philosophical Journal*
*Fraser's Magazine*
*Gentleman's Magazine*
*Literary Gazette*
*Magazine of Natural History*
*Magazine of Popular Science*
*Mining Review*
*Oxford University Calendar*
*Oxford University Magazine*
*Philosophical Magazine*
*Year-book of Science*

AGASSIZ, L., *Recherches sur les poissons fossiles* (5 vols., Neuchâtel, 1833–43).
—— 'On a New Classification of Fishes, and on the Geological Distribution of Fossil Fishes', *Proc. Geol. Soc.* ii (1838), 99–102.
—— 'Upon Glaciers, Moraines, and Erratic Blocks', *Edinburgh New Phil. Journ.* xxiv (1838), 346–83.
—— 'Remarks on Glaciers', *Edinburgh New Phil. Journ.* xxvii (1839), 383–90.
—— *Geologie und Mineralogie* (1838–9), *see under* Buckland.
—— *Etudes sur les glaciers* (2 vols., Neuchâtel, 1840).
—— 'On Glaciers and Boulders in Switzerland', *Report BAAS, 1840,* Trans. Sect., 113–14.
—— 'On the Polished and Striated Surfaces of the Rocks which Form the Beds of Glaciers in the Alps', *Proc. Geol. Soc.* iii (1842), 321–2.
—— 'On Glaciers, and the Evidence of their having once Existed in Scotland, Ireland, and England', *Proc. Geol. Soc.* iii (1842), 327–32.
—— 'On the Succession and Development of Organised Beings at the Surface of the Terrestrial Globe', *Edinburgh New Phil. Journ.* xxxiii (1842), 388–99.
ANON., 'Gisborne's *Natural Theology*', *Quarterly Review* xxi (1819), 41–66.

ANON., *On the Mammoth or Fossil Elephant, Found in the Ice at the Mouth of the River Lena, in Siberia* (London, 1819).

ANON., 'Reflections on the Noachian Deluge, and on Attempts lately made at Oxford, for Connecting the Same with present Geological Appearances', *Phil. Mag. and Journ.* lvi (1820), 10–14.

ANON., 'Mineral and Mosaical Geologies', *Quart. Journ. Sci.* xv (1823), 108–27.

ANON., 'Dr. Daubeny on Volcanoes', *Ann. Phil.* xii (1826), 215–26.

ANON., 'Diary for the Month of February', *London Mag.* x (1828), 360–1.

ANON., 'Eaton's Geology', *Monthly Am. Journ. Geol. Nat. Sci.* i (1831), 82–91.

ANON., 'Discoveries of the Modern Geologists', *Fraser's Mag.* v (1832), 552–66.

ANON., *'A General View of the Geology of Scripture* by George Fairholme', *Athenaeum*, 13 Apr. 1833, 228.

ANON., 'Life in Oxford', *Oxford Univ. Mag.* i (1834), 95–106.

ANON., 'Des traveaux et des résultats de la géologie moderne', *Revue Britannique* xi (1837), 5–28.

ANON., 'The Antediluvians; or, the World Destroyed', *Blackwood's Edinburgh Mag.* xlvi (1839), 119–44.

ANON., 'The Fossil Reptiles of England', *Literary Gazette* 14 Aug. 1841, 513–19.

ANON., *The Caves of the Earth: their Natural History, Features, and Incidents* (London, 1847).

ANSTED, D., *Geology, Introductory, Descriptive, and Practical* (2 vols., London, 1844).

ASHE, T., *Memoirs of Mammoth, and various other Extraordinary and Stupendous Bones* (Liverpool, 1806).

BABBAGE, C., *Reflections on the Decline of Science in England, and on some of its Causes* (London, 1830).

—— *The Ninth Bridgewater Treatise. A Fragment* (London, 1838).

BAKEWELL, F. C., *Natural Evidence of a Future Life, derived from the Properties and Actions of Animate and Inanimate Matter. A Contribution to Natural Theology, designed as a Sequel to the Bridgewater Treatises* (London, 1840).

BAKEWELL, R., *An Introduction to Geology* (London, 1813; 1815; 1828; 1833; 1838).

—— 'Facts and Observations relating to the Theory of the Progressive Development of Organic Life', *Phil. Mag. Ann. Chem.* ix (1831), 33–7.

BAYLE, P., *Dictionnaire historique et critique* (2 vols., Rotterdam, 1695–7).

BEAUMONT, L. E. DE, Researches on some of the Revolutions which have Taken Place on the Surface of the Globe; Presenting various Examples of the Coincidence between the Elevation of Beds in certain Systems of Mountains, and the Sudden Changes which have Produced the Lines

of Demarcation Observable in certain Stages of Sedimentary Deposits',
*Phil. Mag. Ann. Chem.* x (1831), 241–64.

BECHE, H. T. DE LA, 'Notice on the Diluvium of Jamaica', *Ann. Phil.*
x (1825), 54–8.

—— *A Tabular and Proportional View of the Superior (Alluvial and Tertiary),
Supermedial (Secondary Rocks), and Medial Rocks (partly Secondary and partly
Transition Rocks)* (London, 1827).

—— *Geological Notes* (London, 1830).

—— *A Geological Manual* (London, 1831).

—— *Researches in Theoretical Geology* (London, 1834).

—— *The Geological Observer* (London, 1851).

BEST, S., *After Thoughts on Reading Dr. Buckland's Bridgewater Treatise*
(London, 1837).

BICHENO, J., 'On Systems and Methods in Natural History', *Phil. Mag.*
iii (1828), 213–19, 265–71.

BIGELOW, J., 'Documents and Remarks respecting the Sea Serpent', *Am.
Journ. Sci.* ii (1820), 147–64.

BISCHOF, K. G., 'On the Natural History of Volcanos and Earthquakes',
*Edinburgh New Phil. Journ.* xxvi (1839), 25–81, 347–86.

—— 'Farther Reasons against the Chemical Theory of Volcanos',
*Edinburgh New Phil. Journ.* xxx (1841), 14–26.

—— *Physical, Chemical and Geological Researches on the Internal Heat of the Globe*
(London, 1841).

—— 'On the Terrestrial Arrangements Connected with the Appearance
of Man on the Earth', *Edinburgh New Phil. Journ.* xxxvii (1844), 44–62.

BLAINVILLE, H. DE, 'Doubts respecting the Class, Family, and Genus to
which the Fossil Bones found at Stonesfield, and Designated by the
Names *Didelphis Prevostii* and *Did. Bucklandii,* should be Referred', *Mag.
Nat. Hist.* ii (1838), 639–54.

—— 'New Doubts relating to the Supposed *Didelphis* of Stonesfield', *Mag.
Nat. Hist.* iii (1839), 49–57.

BLUMENBACH, J. F., *Beyträge zur Naturgeschichte,* i (Göttingen, 1790; 1806).

BOUBÉE, N., *Géologie élémentaire* (Paris, 1833).

BOUÉ, A., 'Synoptical Table of the Formations of the Crust of the Earth,
and of the Chief Subordinate Masses', *Edinburgh Phil. Journ.* xiii (1825),
130–45.

BOYD, H. S., 'On Cosmogony', *Phil. Mag. and Journ.* 1 (1817), 375–8.

BRANDE, W. T., *A Descriptive Catalogue of the British Specimens Deposited in the
Geological Collection of the Royal Institution* (London, 1816).

—— *Outlines of Geology* (London, 1829).

(BREWSTER, D.), 'Observations on the Decline of Science in England',
*Edinburgh Journ. Sci.* v (1831), 1–16.

—— 'Dr Buckland's *Bridgewater Treatise—Geology and Mineralogy*', *Edinburgh
Review* lxv (1837), 1–39.

BREWSTER, D., *More Worlds than One: the Creed of the Philosopher, and the Hope of the Christian* (London, 1854).

BRODERIP, W. J., 'Observations on the Jaw of a Fossil Mammiferous Animal, found in the Stonesfield Slate', *Zool. Journ.* iii (1828), 408–12.

(BRODERIP, W. J.), 'Agassiz on Fossil Fishes', *Quarterly Review* lv (1836), 433–45.

BRONGNIART, Ad., *Prodrome d'une histoire des végétaux fossiles* (Paris, 1828).

BRONGNIART, Al., 'Notice sur des végétaux fossiles traversant les couches du terrain houiller', *Annales des mines* vi (1821), 359–70.

—— *Tableau des terrains qui composent l'écorce du globe, ou essai sur la structure de la partie connue de la terre* (Paris, 1829).

BRONN, H. G., *Lethaea Geognostica, oder Abbildungen und Beschreibungen der für die Gebirgs-Formationen bezeichnendsten Versteinerungen* (2 vols., Stuttgart, 1835–8).

BROUGHAM, H., *A Discourse of Natural Theology, Showing the Nature of the Evidence and the Advantages of the Study* (London, 1835).

BROWN, J. M., *Reflections on Geology: Suggested by the Perusal of Dr Buckland's Bridgewater Treatise* (London, 1838).

BRYANT, J., *A New System, or, an Analysis of Ancient Mythology* (London, 1774).

BUCH, L. VON, 'Ueber das Fortschreiten der Bildungen in der Natur', *Leopold von Buch's Gesammelte Schriften*, ed. J. Ewald, J. Roth, and H. Eck, ii (Berlin, 1870), 4–12.

BUCKLAND, W., 'Description of a Series of Specimens from the Plastic Clay near Reading, Berks: with Observations on the Formation to which those Beds Belong', *Trans Geol. Soc.* iv (1817), 277–304.

—— *Vindiciae Geologicae; or the Connexion of Geology with Religion Explained* (Oxford, 1820).

—— 'Notice on the Geological Structure of a Part of the Island of Madagascar, Founded on a Collection Transmitted to the Right Honourable the Earl Bathurst, by Governor Farquhar, in the Year 1819; with Observations on some Specimens from the Interior of New South Wales, Collected during Mr. Oxley's Expedition to the River Macquarie, in the Year 1818, and Transmitted also to Earl Bathurst', *Trans. Geol. Soc.* v (1821), 476–81.

—— 'On the Quartz Rock of the Lickey Hill in Worcestershire, and of the Strata immediately Surrounding it; with Considerations on the Evidence of a Recent Deluge Afforded by the Gravel Beds of Warwickshire and Oxfordshire, and the Valley of the Thames from Oxford downwards to London', *Trans. Geol. Soc.* v (1821), 506–44.

—— 'Notice of a Paper laid before the Geological Society on the Structure of the Alps and Adjoining Parts of the Continent, and their Relation to the Secondary and Transition Rocks of England', *Ann. Phil.* i (1821), 450–68.

——— 'Instructions for Conducting Investigations, and Collecting Specimens', *Am. Journ. Sci.* iii (1821), 249–51.

'Opinion of Professor Buckland of the University of Oxford, Respecting certain Features of American Geology', *Am. Journ. Sci.* iv (1822), 185–6.

——— 'Account of an Assemblage of Fossil Teeth and Bones of Elephant, Rhinoceros, Hippopotamus, Bear, Tiger, and Hyaena, and Sixteen other Animals; Discovered in a Cave at Kirkdale, Yorkshire, in the Year 1821: with a Comparative View of Five Similar Caverns in Various Parts of England, and Others on the Continent', *Abstracts of the Papers, Phil. Trans.* ii (1823), 165–7; *Phil. Trans.* cxii (1822), 171–236.

——— *Reliquiae Diluvianae; or, Observations on the Organic Remains Contained in Caves, Fissures, and Diluvial Gravel, and on other Geological Phenomena, Attesting the Action of an Universal Deluge* (London, 1823; 1824).

——— 'On the Excavation of Valleys by Diluvial Action, as Illustrated by a Succession of Valleys which Intersect the South Coast of Dorsetshire and Devonshire', *Trans. Geol. Soc.* i (1824), 95–102.

——— 'Notice on the Megalosaurus, or Great Fossil Lizard of Stonesfield', *Trans. Geol. Soc.* i (1824), 390–6.

——— 'Reply to some Observations in Dr. Fleming's Remarks on the Distribution of British Animals', *Edinburgh Phil. Journ.* xii (1825), 304–19.

'Letter of Professor Buckland to Professor Jameson, and of Captain Sykes to Professor Buckland, on the Interior of the Dens of Living Hyaenas', *Edinburgh New Phil. Journ.* ii (1827), 377–80.

——— 'On the Formation of the Valley of Kingsclere and other Valleys, by the Elevation of the Strata that Enclose them; and on the Evidences of the Original Continuity of the Basins of London and Hampshire', *Trans. Geol. Soc.* ii (1829), 119–130.

——— 'Geological Account of a Series of Animal and Vegetable Remains and of Rocks, Collected by J. Crawfurd, Esq. on a Voyage up the Irawadi to Ava, in 1826 and 1827', *Trans. Geol. Soc.* ii (1829), 377–92.

——— 'On the Cycadeoideae, a Family of Fossil Plants Found in the Oolite Quarries of the Isle of Portland', *Trans. Geol. Soc.* ii (1829), 395–401.

——— 'Antediluvian Human Remains', *Am. Journ. Sci.* xviii (1830), 393–4.

——— 'Bones in Caves, etc.', *Monthly Am. Journ. Geol. Nat. Sci.* i (1831), 278–80.

——— 'On the Fossil Remains of the Megatherium', *Report BAAS, 1831. 1832,* 104–7.

——— 'On the Occurrence of the Remains of Elephants, and other Quadrupeds, in the Cliffs of Frozen Mud, in Eschscholtz Bay, within Beering's Strait, and in other Distant Parts of the Shores of the Arctic Seas', F. W. Beechey, *Narrative of a Voyage to the Pacific and Beering's Strait* (London, 1831), 593–612.

——— 'Observations on the Bones of Hyaenas and Other Animals in the

Cavern of Lunel near Montpelier, and in the Adjacent Strata of Marine Formations', *Proc. Geol. Soc.* i (1834), 3–6.

—— 'An Account of the Discovery of a Number of Fossil Bones of Bears, in the Grotto of Osselles, or Quingey, near Besançon in France', *Proc. Geol. Soc.* i (1834), 21–2.

—— 'On the Discovery of a New Species of Pterodactyle; and also of the Faeces of the Ichthyosaurus; and of a Black Substance Resembling Sepia, or Indian Ink, in the Lias at Lyme Regis', *Proc. Geol. Soc.* i (1834), 96–8.

—— 'On the Discovery of the Bones of the Iguanodon, and other Large Reptiles, in the Isle of Wight and Isle of Purbeck', *Proc. Geol. Soc.* i (1834), 159–60.

—— 'On the Discovery of a New Species of Pterodactyle in the Lias at Lyme Regis', *Trans. Geol. Soc.* iii (1835), 217–22.

—— 'On the Discovery of Coprolites, or Fossil Faeces, in the Lias at Lyme Regis, and in other Formations', *Trans. Geol. Soc.* iii (1835), 223–36.

—— *Geology and Mineralogy Considered with Reference to Natural Theology* (2 vols., London, 1836; 1837; 1858; 1869).

—— 'On the Adaptation of the Structure of the Sloths to their peculiar Mode of Life', *Trans. Linnean Soc.* xvii (1837), 17–27.

—— 'A Notice on the Fossil Beaks of Four Extinct Species of Fishes, referrible to the Genus Chimaera, which Occur in the Oolitic and Cretaceous Formations of England', *Proc. Geol. Soc.* ii (1838), 205–6.

—— 'On the Discovery of Fossil Fishes in the Bagshot Sands at Goldworth Hill, Four Miles North of Guildford', *Proc. Geol. Soc.* ii (1838), 687–8.

—— 'An Account of the Footsteps of the Cheirotherium and Five or Six Smaller Animals in the Stone Quarries of Storeton Hill, near Liverpool, Communicated by the Natural History Society of Liverpool', *Report BAAS, 1838,* Trans. Sect. 85.

—— *Geologie und Mineralogie in Beziehung zur natürlichen Theologie,* translated and annotated by L. Agassiz (2 vols., Neuchâtel, 1838–9).

—— *An Inquiry whether the Sentence of Death pronounced at the Fall of Man Included the whole Animal Creation, or was Restricted to the Human Race* (London, 1839).

—— Anniversary address to the Geological Society, 21 Feb. 1840, *Proc. Geol. Soc.* iii (1842), 210–67.

—— 'Memoir on the Evidences of Glaciers in Scotland and the North of England', *Proc. Geol. Soc.* iii (1842), 332–7, 345–8.

—— Anniversary address to the Geological Society, 19 Feb. 1841, *Proc. Geol. Soc.* iii (1842), 469–540.

—— 'On the Glacia-diluvial Phaenomena in Snowdonia and the adjacent Parts of North Wales', *Proc. Geol. Soc.* iii (1842), 579–84.

—— 'On Fossil Impressions of Rain and Ripple Marks', *Abstracts Proc. Ashmolean Soc.* i (1844), no. xvi, 5–7.

—— 'On the Glacial Theory', *Abstracts Proc. Ashmolean Soc.* i (1844), no. xvii, 22–4.

—— 'On the Agency of Animalcules in the Formation of Limestone', *Abstracts Proc. Ashmolean Soc.* i (1844), no. xvii, 35–9.

—— 'Further Remarks on the Glacial Theory', *Abstracts Proc. Ashmolean Soc.* i (1844), no. xviii, 17–19.

—— *Address to the Mayor and Members of the Artesian Well Committee, of Southampton, July 27, 1844* (Southampton, 1844).

(BUCKLAND, W.), 'Agriculture', *Quarterly Review* lxxiii (1844), 477–509.

'Report of Dr. Buckland's Observations, at the Canterbury Meeting of the British Archaeological Association', *Report of the Proceedings of the British Archaeological Association, 1844, 1845,* 106–13.

BUCKLAND, W., *A sermon Preached in Westminster Abbey, on Easter Sunday Evening, April 23, 1848, on the Occasion of the Re-opening of the Choir, and the Application of the Transepts to the Reception of the Congregation* (London, 1848).

—— *A Sermon Preached in Westminster Abbey, on the 15th Day of November, 1849; being the Day of Thanksgiving to God for the Removal of the Cholera* (London, 1849).

*Catalogue of the Theological and Miscellaneous Library of the Late Very Reverend Dr. Buckland, Dean of Westminster* (London, 1857).

*A Catalogue of the Valuable Scientific Library of the Late Very Rev. Dr. Buckland, Dean of Westminster* (London, 1857).

BUCKLAND, W., and H. T. DE LA BECHE, 'On the Geology of the Neighbourhood of Weymouth and the adjacent Parts of the Coast of Dorset', *Trans. Geol. Soc.* iv (1836), 1–46.

BUCKLAND, W., and W. D. CONYBEARE, 'Observations on the South-Western Coal District of England', *Trans. Geol. Soc.* i (1824), 210–316.

BUFFON, G. L. LECLERQ, COMTE DE, *Les Époques de la nature* (3rd edn., 2 vols., Paris, 1790).

(BUGG, G.), *Scriptural Geology; or, Geological Phenomena, Consistent only with the Literal Interpretation of the Sacred Scriptures, upon the Subjects of the Creation and Deluge; in Answer to an 'Essay on the Theory of the Earth' by M. Cuvier, Perpetual Secretary of the French Institute, etc. and to Dr. Buckland's Theory of the Caves, as Delineated in his 'Reliquiae Diluvianae', etc.* (2 vols., London, 1826–7).

(BULWER-LYTTON, E. G.), *England and the English* (2 vols., New York, 1833).

BUTLER, J., *The Analogy of Religion, Natural and Revealed, to the Constitution and Course of Nature* (Oxford, 1844).

*The Poetical Works of Lord Byron,* ed. E. H. Coleridge (London, 1905).

CARROL, L., *The Annotated Alice,* ed. M. Gardner, (Penguin, 1979).

CATCOTT, A., *A Treatise on the Deluge* (2nd edn., London, 1768).

CHALMERS, T., *The Evidence and Authority of the Christian Revelation* (Edinburgh, 1814).

—— *On Political Economy in Connexion with the Moral State and Moral Prospects of Society* (Glasgow, 1832).

—— *On the Power, Wisdom and Goodness of God as Manifested in the Adaptation of External Nature to the Moral and Intellectual Constitution of Man* (2 vols., London, 1833).

(CHAMBERS, R.), *Vestiges of the Natural History of Creation*, facsimile of the London, 1844 edn., introduction by G. de Beer (New York, 1969).

CHARPENTIER, J. DE, 'Sur un tronc d'arbre fossile', *Bibliotheque Universelle* ix (1818), 254–8.

CHRISTISON, R., 'Notice of Fossil Trees recently Discovered in Graigleith-Quarry, near Edinburgh', *Trans. Roy. Soc. Edinburgh* xxvii (1876), 203–21.

CLINTON, H. F., *Fasti Hellenici. The Civil and Literary Chronology of Greece from the Earliest Account to the Death of Augustus*, i (Oxford, 1834).

COCKBURN, W., *A Letter to Professor Buckland, Concerning the Origin of the World* (London, 1838).

—— *A Remonstrance, Addressed to His Grace the Duke of Northumberland, upon the Dangers of Peripatetic Philosophy* (London, 1838).

—— *The Creation of the World. Addressed to R. J. Murchison, Esq. and Dedicated to the Geological Society* (London, 1840).

—— *The Bible Defended against the British Association* (London, 1844).

—— *A New System of Geology; Dedicated to Professor Sedgwick* (London, 1849).

COLE, H., *Popular Geology Subversive of Divine Revelation! A Letter to the Rev. Adam Sedgwick* (London, 1834).

CONYBEARE, W. D., 'Memoir Illustrative of a general Geological Map of the principal Mountain Chains of Europe', *Ann. Phil.* v (1823), 1–16, 135–49, 210–18, 278–89, 356–9; vi (1823), 214–19.

—— 'Additional Notices on the Fossil Genera Ichthyosaurus and Plesiosaurus', *Trans Geol. Soc.* i (1824), 103–23.

—— 'Answer to Dr Fleming's View of the Evidence from the Animal Kingdom, as to the Former Temperature of the Northern Regions', *Edinburgh New Phil. Journ.* Apr. to Oct. 1829, 142–52.

—— 'Letter on Mr. Lyell's "Principles of Geology"', *Phil. Mag. Ann., Chem.* viii (1830), 215–19.

—— 'An Examination of those Phaenomena of Geology, which Seem to Bear most Directly on Theoretical Speculations', *Phil. Mag. Ann. Chem.* viii (1830), 359–62, 401–6; ix (1831), 19–23, 111–17, 188–97, 258–70.

—— *Inaugural Address on the Application of Classical and Scientific Education to Theology; and on the Evidences of Natural and Revealed Religion* (London, 1831).

—— 'Report on the Progress, Actual State, and Ulterior Prospects of Geological Science', *Report BAAS, 1831, 1832*, 365–414.

—— 'On the Valley of the Thames', *Proc. Geol. Soc.* i (1834), 145–9.

CONYBEARE, W. D., and H. T. DE LA BECHE, 'Notice of the Discovery of a New Fossil Animal, Forming a Link between the Ichthyosaurus and Crocodile, together with general Remarks on the Osteology of the Ichthyosaurus', *Trans. Geol. Soc.* v (1821), 559–94.

CONYBEARE, W. D., and W. PHILLIPS, *Outlines of the Geology of England and Wales, with an Introductory Compendium of the General Principles of that Science, and Comparative Views of the Structure of Foreign Countries,* i (London, 1822).

(CONYBEARE, W. J.), 'Church Parties', *Edinburgh Review* xcviii (1853), 273–342.

(COPLESTON, E.), *A Reply to the Calumnies of the Edinburgh Review against Oxford. Containing an Account of Studies pursued in that University* (Oxford, 1810).

—— *A Second Reply to the Edinburgh Review* (Oxford, 1810).

—— *A Third Reply to the Edinburgh Review* (Oxford, 1811).

—— 'Buckland—*Reliquiae Diluvianae*', *Quarterly Review* xxix (1823), 138–65.

CORDIER, L., 'Essai sur la température de l'intérieur de la terre', *Mém. Acad. Sci.* vii (1827), 473–556.

—— 'On the Temperature of the Interior of the Earth', *Edinburgh New Phil. Journ.,* Oct. 1827 to Apr. 1828, 273–90.

—— 'Examination of the Experiments hitherto Published on Subterranean Temperature, together with Experiments and Inquiries relative to this Examination', *Edinburgh New Phil. Journ.* Apr. to Sept. 1828, 277–91; Oct. 1828 to Mar. 1829, 33–45.

COWPER, W., *The Task. A Poem in Six Books* (London, 1883).

CRAWFURD, J., *Journal of an Embassy from the Governor General of India to the Court of Ava,* ii (2nd edn., London, 1834).

CRICHTON, A., 'On the Climate of the Antediluvian World, and its Independence of Solar Influence; and on the Formation of Granite', *Ann. Phil.* ix (1825), 97–108, 207–17.

CROLY, G., *The Apocalypse of St. John, or Prophecy of the Rise, Progress, and Fall of the Church of Rome* (London, 1827).

—— *Divine Providence; or, the Three Cycles of Revelation, showing the Parallelism of the Patriarchal, Jewish, and Christian Dispensations. Being a New Evidence of the Divine Origin of Christianity* (London, 1834).

(CROLY, G.), 'The World We Live in', *Blackwood's Edinburgh Mag.* xlii (1837), 673–92.

CUMBERLAND, G., *Reliquiae Conservatae, from the Primitive Materials of our Present Globe, with Popular Descriptions of the Prominent Characters of some Remarkable Fossil Encrinites, and their Connecting Links* (Bristol, 1826).

CUNINGHAME, W., *The Fulness of the Times: being an Analysis of the Chronology of the Greek Text of the Seventy: showing that it Rests on the Basis of Exact Science, and Comprehends various Parallel Streams of Time, Arranged in Great Periods of Jubilees and Astronomical Cycles, which Connect the Eras of History and Prophecy with the Remotest Antediluvian Ages, and Demonstrate the Divine Origin of the Hebrew Dispensation. Proving also that we are Approaching a Great Crisis in the Sacred Chronology, which Indicates the Nearness of the End* (London, 1836).

—— *A Synopsis of Chronology from the Era of Creation, according to the Septuagint, to the Year 1837; with a Discourse on the Astronomical Principles of the Scriptural Times, showing that they Comprehend a Complex Harmony of Deeply Scientific Order and Arrangement, demonstrating their Exact Truth, and evincing that their Author is the Omniscient Creator* (London, 1837).

—— *The Scientific Chronology of the Year 1839, a Sign of the Near Approach of the Kingdom of God* (London, 1839).

—— *The Season of the End; being a View of the Scientific Times of the Year 1840* (london, 1841).

—— *The Certain Truth, the Science, and the Authority of the Scriptural Chronology* (London, 1849).

—— *A Letter to the Right Honourable Lord Ashley, President of the London Society for Promoting Christianity amongst the Jews, on the Necessity of Immediate Measures for the Jewish Colonization of Palestine* (London, 1849).

CUVIER, G., *Essay on the Theory of the Earth. Translated by Robert Kerr, with Mineralogical Notes by Professor Jameson* (Edinburgh, 1813).

—— *Le Règne animal distribué d'apres son organisation* (4 vols., Paris, 1817).

—— 'Historical Eloge of Abraham Gottlob Werner', *Edinburgh Phil. Journ.* iv (1821), 1–16.

—— *Recherches sur les ossemens fossiles, où l'on rétablit les caractères de plusieurs animaux dont les révolutions du globe ont détruit les espèces,* iv (Paris, 1823).

—— *Discours sur les révolutions de la surface du globe* (Paris, 1830).

CUVIER, G., and A. BRONGNIART, 'Essai sur la géographie minéralogique des environs de Paris', *Annales de Muséum d'Histoire Naturelle* xi (1808), 293–326.

DARWIN, C. R., 'Observations on the Parallel Roads of Glen Roy', *Phil. Trans.* cxxix (1839), 39–81.

—— *The Origin of Species by Means of Natural Selection, or the Preservation of Favoured Races in the Struggle for Life* (London, 1859).

DAUBENY, C. G. B., *Inaugural Lecture on the Study of Chemistry* (Oxford, 1823).

—— *A Description of Active and Extinct Volcanos, with Remarks on their Origin, their Chemical Phaenomena, and the Character of their Products, as Determined by the Condition of the Earth during the Period of their Formation* (London, 1826).

—— 'On the Diluvial Theory, and on the Origin of the Valleys of Auvergne', *Edinburgh New Phil. Journ.* Oct. 1830 to Apr. 1831, 201–29.

(DAUBENY, C. G. B.), 'Apology for British Science', *London Literary Gazette,* 7 Dec. 1833, 769–71, 789–92.

DAUBENY, C. G. B., *An Inaugural Lecture on the Study of Botany* (Oxford, 1834).

——'To the Members of Convocation', 24 Feb. 1839, 11 Mar. 1839, *Papers Relating to the Proceedings of the University* (1839).

——'Reply to Professor Bischof's Objections to the Chemical Theory of Volcanos', *Edinburgh New Phil. Journ.* xxvi (1839), 291–9; xxvii (1839), 158–60.

——*Brief Remarks on the Correlation of the Natural Sciences* (Oxford, 1848).

(DAUBENY, C. G. B.), *A Dream of the New Museum* (Oxford, 1855).

DAUBENY, C. G. B., *Fugitive Poems Connected with Natural History and Physical Science* (Oxford, 1869).

DAVY, H., *Six Discourses Delivered before the Royal Society at their Anniversary Meetings, on the Award of the Royal and Copley Medals* (London, 1827).

——*Consolations in Travel: or the Last Days of a Philosopher* (London, 1830).

—— 'On the Formation of the Earth', *Edinburgh New Phil. Journ.* Jan. to Apr. 1830, 320–2.

DELUC, G. A., 'Lettre aux rédacteurs', *Bibliotheque Britannique,* xviii (1801), 272–98.

DELUC, J. A., *An Elementary Treatise on Geology: Determining Fundamental Points in that Science, and Containing an Examination of some Modern Geological Systems, and Particularly of the Huttonian Theory of the Earth. Translated from the French Manuscript, by the Rev. Henry de la Fite* (London, 1809).

DELUC, J. A. (the younger), 'On the Glaciers of the Alps', *Edinburgh New Phil. Journ.* xxviii (1840), 15–20.

DERHAM, W., *Physico-Theology: or, a Demonstration of the Being and Attributes of God, from his Works of Creation* (London, 1754).

DESNOYER, J., 'Proofs that the Human Bones and Works of Art found in Caves in the South of France, are more Recent than the Antediluvian Bones in these Caves', *Edinburgh New Phil. Journ.* xvi (1834), 302–10.

DICKENS, C., *Bleak House* (Oxford, 1978).

(DUNCAN, J. S.), *Botano-Theology, an Arranged Compendium, Chiefly from Smith, Keith, and Thompson* (Oxford, 1825).

DUNCAN, J. S., 'A Summary Review of the Authorities on which Naturalists are Justified in Believing that the Dodo, *Didus ineptus, Linn.,* was a Bird Existing in the Isle of France, or the Neighbouring Islands, until a Recent Period', *Zool. Journ.* iii (1828), 554–66.

——*Analogies of Organized Beings* (Oxford, 1831).

——'On Sea Serpents', *Abstracts Proc. Ashmolean Soc.* i (1844), no. iii, 4–5.

(DUNCAN, P. B.), *A Catalogue to the Ashmolean Museum* (Oxford, 1836).

DUNCAN, P. B., *Literary Conglomerate; or, a Combination of various Thoughts and Facts on various Subjects* (Oxford, 1839).

——*Motives of War* (London, 1844).

EASTMEAD, W., *Historia Rievallensis: Containing the History of Kirby Moorside, and an Account of the most Important Places in the Vicinity; together with Brief Notices of the more Remote or less Important Ones. To Which is Prefixed a*

*Dissertation on the Animal Remains, and other Curious Phenomena, in the Recently Discovered Cave at Kirkdale* (Thirsk, 1824).

EATON, A., 'Notices respecting Diluvial Deposits in the State of New-York and Elsewhere', *Am. Journ. Sci.* xii (1827), 17–20.

——'Geological Nomenclature, Classes of Rocks, etc.', *Am. Journ. Sci.* xiv (1828), 145–59, 359–68.

EHRENBERG, C. G., 'On the Calcareous and Siliceous Microscopic Animals which Form the Chief Component Parts of Cretaceous Rocks', *Edinburgh New Phil. Journ.* xxviii (1840), 161–6.

ELIOT, G., *The Mill on the Floss* (Penguin edn., Aylesbury, 1979).

ELLIOTT, E. B., *Horae Apocalypticae; or a Commentary on the Apocalypse Critical and Historical; Including also an Examination of the Chief Prophecies of Daniel* (3 vols., London, 1844).

ESMARK, J., 'Remarks tending to Explain the Geological History of the Earth', *Edinburgh New Phil. Journ.* Oct. 1826 to Apr. 1827, 107–21.

ESPER, J. F., *Ausführliche Nachricht von neuentdeckten Zoolithen unbekannter vierfüssiger Tiere*, facsimile of the Nürnberg 1774 edn., introduced by A. Geus (Wiesbaden, 1978).

FABER, G. S., *Horae Mosaicae; or a View of the Mosaical Records, with Respect to their Coincidence with Profane Antiquity; their Internal Credibility; and their Connection with Christianity* (Oxford, 1801).

——*The Difficulties of Infidelity* (London, 1824).

FAIRHOLME, G., *General View of the Geology of Scripture, in which the Unerring Truth of the Inspired Narrative of the Early Events in the World is Exhibited, and Distinctly Proved, by the Corroborative Testimony of Physical Facts, on Every Part of the Earth's Surface* (London, 1833).

——'Some Observations on the Nature of Coal, and on the Manner in which the various Strata of the Coal-measures Must probably have been Deposited', *London and Edinburgh Phil. Mag. and Journ. Sci.* iii (1833), 245–52.

(FAIRHOLME, G.), 'A Layman on Scriptural Geology: with Observations Thereon', *Christian Observer*, 1834, 479–92.

FAIRHOLME, G., *Positions géologiques, en vérification directe de la chronologie de la Bible* (Munich, 1834).

——*New and Conclusive Physical Demonstrations, both of the Fact and Period of the Mosaic Deluge, and of its having been the only Event of the Kind that has ever Occurred upon the Earth* (London, 1837).

FAREY, J., 'Mr. Smith's Geological Claims Stated', *Phil. Mag. and Journ.* li (1818), 173–80.

——'Free Remarks on Mr. Greenough's Gelogical Map, lately Published under the Direction of the Geological Society of London', *Phil. Mag. and Journ.* lv (1820), 379–83.

FEATHERSTONHAUGH, G. W., 'Bone Caves in New Holland', *Monthly Am. Journ. Geol. Nat. Sci.* i (1831), 47.

(FEATHERSTONHAUGH, G. W.), 'On the Order of Succession of the Rocks Composing the Crust of the Earth', *Monthly Am. Journ. Geol. Nat. Sci.* i (1832), 337–47.

FEATHERSTONHAUGH, G. W., 'On the Series of Rocks in the United States', *Proc. Geol. Soc.* i (1834), 91–3.

FERGUS, H., *The Testimony of Nature and Revelation to the Being, Perfections, and Government of God* (Edinburgh, 1833).

FIGUIER, L., *La Terre avant le déluge* (Paris, 1863).

(FITTON, W. H.), 'Transactions of the Geological Society, Vol. III', *Edinburgh Review* xxix (1818), 70–94.

—— 'Geology of England', *Edinburgh Review* xxix (1818), 310–37.

—— 'Geology of the Deluge', *Edinburgh Review* xxxix (1824), 196–234.

FITTON, W. H., 'On the Strata from whence the Fossil Described in the Preceding Notice was Obtained', *Zool. Journ.* iii (1828), 412–18.

F(ITTON?), 'MacCulloch's *System of Geology*', *Edinburgh Journ. Sci.* v (1831), 358–75.

FITTON, W. H., 'Notes on the History of English Geology', *London and Edinburgh Phil. Mag. and Journ. Sci.* i (1832), 147–60, 268–75, 442–50; ii (1833), 37–57.

——Anniversary address to the Geological Society, 20 Feb. 1829, *Proc. Geol. Soc.* i (1834), 112–34.

——'Observations on Some of the Strata between the Chalk and the Oxford Oolite, in the South-east of England', *Trans. Geol. Soc.* iv (1836), 103–389.

FLEMING, J., *The Philosophy of Zoology; or, a General View of the Structure, Functions, and Classification of Animals* (2 vols., Edinburgh, 1822).

——'On the Revolutions which have taken Place in the Animal Kingdom, as these are Indicated by Geognosy', *Edinburgh Phil. Journ.* viii (1823), 110–22.

——'Remarks Illustrative of the Influence of Society on the Distribution of British Animals', *Edinburgh Phil. Journ.* xi (1824), 287–305.

——'Remarks on the Modern Strata', *Edinburgh Phil. Journ.* xii (1825), 116–27.

——'The Geological Deluge, as interpreted by Baron Cuvier and Professor Buckland, inconsistent with the Testimony of Moses and the Phenomena of Nature', *Edinburgh Phil. Journ.* xiv (1826), 205–39.

——'On the Value of the Evidence from the Animal Kingdom, tending to Prove that the Arctic Regions formerly Enjoyed a Milder Climate than at Present', *Edinburgh New Phil. Journ.* Oct. 1828 to Mar. 1829, 277–86.

(FLEMING, J.), 'Systems and Methods in Natural History', *Quarterly Review* xli (1829), 302–27.

FORBES, J. D., 'An Attempt to Explain the Leading Phenomena of Glaciers', *Edinburgh New Phil. Journ.* xxxv (1843), 221–52.

FOX, R. W., 'Report on Some Observations on Subterranean Temperature', *Report BAAS, 1840,* 309–19.

FRAAS, O., *Vor der Sündfluth! Eine Geschichte der Urwelt* (Stuttgart, 1866).

GEOFFROY SAINT-HILAIRE, É., *Recherches sur de grands sauriens trouvés a l'état fossile vers les confins maritimes de la Basse Normandie, attribués d'abord au crocodile, puis déterminés sous les noms de téléosaurus et sténéosaurus* (Paris, 1831).

GISBORNE, T., *The Testimony of Natural Theology to Christianity* (London, 1818).

GISSING, G., *Born in Exile*, ed. P. Coustillas (Hassocks, Sussex, 1978).

GOLDFUSS, G. A., *Petrefacta Germaniae* (3 vols., Düsseldorf, 1826–44).

GÖPPERT, H. R., *Abhandlung, eingesandt als Antwort auf die Preisfrage: 'Man suche durch genaue Untersuchungen darzuthun, ob die Steinkohlenlager aus Pflanzen entstanden sind, welche an den Stellen, wo jene gefunden werden, wuchsen; oder ob diese Pflanzen an anderen Orten lebten, und nach den Stellen, wo sich die Steinkohlenlager befinden, hingeführt wurden?'* (Haarlem, 1848).

GOSSE, P. H., *Omphalos: an Attempt to Untie the Geological Knot* (London, 1857).

——*The Romance of Natural History* (London, 1860).

(GRANT, R. E.), 'Observations on the Nature and Importance of Geology', *Edinburgh New Phil. Journ.* Apr. to Oct. 1826, 293–302.

GRAY, J., *The Earth's Antiquity in Harmony with the Mosaic Record of Creation* (London 1849; 1951).

GREENOUGH, G. B., *A Critical Examination of the First Principles of Geology; in a Series of Essays* (London, 1819).

——Anniversary address to the Geological Society, 21 Feb. 1834, *Proc. Geol. Soc.* ii (1838), 42–70.

HACK, M., *Geological Sketches, and Glimpses of the Ancient Earth* (London, 1832).

HALES, W., *A New Analysis of Chronology and Geography, History and Prophecy: in which their Elements are Attempted to be Explained, Harmonized, and Vindicated, upon Scriptural and Scientific Principles; Tending to Remove the Imperfection and Discordance of Preceding Systems, and to Obviate the Cavils of Sceptics, Jews, and Infidels* (2nd edn., 4 vols., London, 1830).

HALL, J., 'On the Revolutions of the Earth's Surface', *Trans. Roy. Soc. Edinburgh* vii (1815), 139–212.

(HAMILTON, W.), 'On the State of the English Universities with more especial reference to Oxford', *Edinburgh Review* liii (1831), 384–427; liv (1831), 478–504.

——'Patronage of Universities', *Edinburgh Review* lix (1834), 196–227.

——'On the Right of Dissenters to Admission into the English Universities', *Edinburgh Review* lx (1834), 202–30.

——'The Universities and the Dissenters', *Edinburgh Review* lx (1835), 422–45.

HAMPDEN, R. D., *An Essay on the Philosophical Evidence of Christianity; or, the Credibility obtained to a Scriptural Revelation, from its Coincidence with the Facts of Nature* (London, 1827).

HAWKINS, T., *Memoirs of Ichthyosauri and Plesiosauri, Extinct Monsters of the Ancient Earth* (London, 1834).
——*The Book of the Great Sea Dragons* (London, 1840).
HENSLOW, J. S., 'On the Deluge', *Ann. Phil.* vi (1823), 344–8.
HERDER, J. G., *Ideen zur Philosophie der Geschichte der Menschheit,* i (Riga, 1784).
HERSCHEL, J. F. W., *Preliminary Discourse on the Study of Natural Philosophy* (London, 1830).
——'On the Astronomical Causes which May Influence Geological Phaenomena', *Trans. Geol. Soc.* iii (1835), 293–9.
HIGGINS, W. M., *The Mosaical and Mineral Geologies Illustrated and Compared* (London, 1832).
——*The Book of Geology* (London, 1842).
HITCHCOCK, E., 'Description of the Foot Marks of Birds (Ornitichnites) on New Red Sandstone in Massachusetts', *Am. Journ. Sci.* xxix (1836), 307–40.
——*The Religion of Geology and its Connected Sciences* (London, 1851).
HOME, E., 'Some Account of the Fossil Remains of an Animal more nearly Allied to Fishes than any of the other Classes of Animals', *Phil. Trans.* civ (1814), 571–7.
——'An Account of some Fossil Remains of the Rhinoceros, Discovered by Mr. Whitby, in a Cavern Inclosed in the Lime-stone Rock, from which he is Forming the Break-water at Plymouth', *Abstracts of the Papers, Phil. Trans.* ii (1823), 66–7; *Phil. Trans.* cvii (1817), 176–82.
HOPKINS, W., 'Researches in Physical Geology', *Trans. Cambridge Phil. Soc.* vi (1838), 1–84.
——'On the Elevation and Denudation of the District of the Lakes of Cumberland and Westmoreland', *Proc. Geol. Soc.* iii (1842), 757–66.
HUMBOLDT, A. VON, *A Geognostical Essay on the Superposition of Rocks, in both Hemispheres* (London, 1823).
HUMBOLDT, A. VON, 'On Rock Formations', *Edinburgh Phil. Journ.* x (1834), 40–53, 224–39.
——*Kosmos. Entwurf einer physischen Weltbeschreibung* (5 vols., Stuttgart and Tübingen, 1845–62).
HUME, D., *Dialogues Concerning Natural Religion,* ed. N. K. Smith (London, 1947).
HUTTON, J., 'Theory of the Earth; or an Investigation of the Laws discernible in the Composition, Dissolution and Restoration of Land upon the Globe', *Trans. Roy. Soc. Edinburgh* i (1788), 209–304.
JACOB, W. S., *A Few More Words on the Plurality of Worlds* (London, 1855).
JAMESON, R., *System of Mineralogy, Comprehending Oryctognosy, Geognosy, Mineralogical Chemistry, Mineralogical Geography, and Economical Mineralogy,* iii (Edinburgh, 1808).

JOHNSONE, F. DE, *A Vindication of the Book of Genesis. Addressed to the Rev. William Buckland* (London, 1838).

KENNEDY, J., *A New Method of Stating and Explaining the Scripture Chronology, upon Mosaic Astronomical Principles, Mediums and Data, as Laid down in the Pentateuch* (London, 1751).

KIDD, J., *A Geological Essay on the Imperfect Evidence in Support of a Theory of the Earth, deducible either from its General Structure or from the Changes Produced on its Surface by the Operation of Existing Causes* (Oxford, 1815).

——*An Answer to a Charge against the English Universities Contained in the Supplement to the Edinburgh Encyclopaedia* (Oxford, 1818).

—— *An Introductory Lecture to a Course in Comparative Anatomy, Illustrative of Paley's Natural Theology* (Oxford, 1824).

——*On the Adaptation of External Nature to the Physical Condition of Man; Principally with Reference to the Supply of his Wants and the Exercise of his Intellectual Faculties* (London, 1833).

KIRBY, W., *On the Power Wisdom and Goodness of God as Manifested in the Creation of Animals and in their History Habits and Instincts* (2 vols., London, 1835).

KIRWAN, R., *Geological Essays* (London, 1799).

KNOX, R., 'Notice Relative to the Habits of the Hyaena of Southern Africa', *Mem. Wernerian Nat. Hist. Soc.* iv (1822), 383–5.

(KNIGHT, R. P.), 'The Oxford Edition of Strabo', *Edinburgh Review* xiv (1809), 429–41.

(KNIGHT, R. P., J. PLAYFAIR, and S. SMITH), 'Calumnies against Oxford', *Edinburgh Review* xvi (1810), 158–87.

KÖNIG, C., 'On a Fossil Human Skeleton from Guadeloupe', *Abstracts of the Papers, Phil. Trans.* i (1832), 487–9.

LANG, 'Account of the Discovery of Bone Caves in Wellington Valley, about 210 Miles West from Sydney in New Holland', *Edinburgh New Phil. Journ.* Oct. 1830 to Apr. 1831, 364–71.

LEIBNIZ, G. W. VON, *Protogea, sive de Prima Facie Telluris et Antiquissimae Historiae Vestigiis in Ipsis Naturae Monumentis Dissertatio* (Göttingen, 1749).

LINDLEY, J., and W. HUTTON, *The Fossil Flora of Great Britain; or, Figures and Descriptions of the Vegetable Remains found in a Fossil State in this Country,* i (London, 1831).

LOGAN, W. E., 'On the Characters of the Beds of Clay immediately below the Coal Seams of South Wales, and on the Occurrence of Boulders of Coal in the Pennant Grit of that District', *Trans. Geol. Soc.* vi (1842), 491–7.

LOUDON, J. C., 'Preface', *Mag. Nat. Hist.* vii (1834), iv.

(LOUDON, J. C.), 'Introduction to Geology. Geological Systems of Arrangement', *Mag. Nat. Hist.* iii (1830), 62–78.

(LYELL, C.), '*Transactions of the Geological Society*', *Quarterly Review* xxxiv (1826), 507–40.

——'State of the Universities', *Quarterly Review* xxxvi (1827), 216–68.

——'Scrope's *Geology of Central France*', *Quarterly Review* xxxvi (1827), 437–83.

LYELL, C., *Principles of Geology, being an Attempt to Explain the Former Changes of the Earth's Surface, by Reference to Causes now in Operation* (3 vols., London, 1830–3).

—— 'Reply to a Note in the Rev. Mr. Conybeare's Paper entitled "An Examination of those Phaenomena of Geology, which Seem to Bear most Directly on Theoretical Speculations" ', *Phil. Mag. Ann. Chem.* ix (1831), 1–3.

—— 'On the Proofs of a Gradual Rising of the Land in Certain Parts of Sweden', *Phil. Trans.* cxxv (1835), 1–38.

—— Anniversary address to the Geological Society, 19 Feb. 1836, *Proc. Geol. Soc.* ii (1838), 357–90.

—— Anniversary address to the Geological Society, 17 Feb. 1837, *Proc. Geol. Soc.* ii (1838), 479–523,

—— *Elements of Geology* (London, 1838; 1841; 1865).

—— 'On the Geological Evidence of the Former Existence of Glaciers in Forfarshire', *Proc. Geol. Soc.* iii (1842), 337–45.

(LYELL, C.), 'The Theory of Successive Development in the Scale of Being both Animal and Vegetable, from the Earliest Periods to our own Time, as Deduced from Palaeontological Evidence', *Edinburgh New Phil. Journ.* li (1851), 1–31, 213–26.

LYELL, C., and R. I. MURCHISON, 'On the Excavation of Valleys, as illustrated by the Volcanic Rocks of Central France', *Proc. Geol. Soc.* i (1834), 89–91.

—— 'On the Excavation of Valleys, as illustrated by the Volcanic Rocks of Central France', *Edinburgh New Phil. Journ.* Apr. to Oct. 1829, 15–48.

McCOMBIE, W., *Moral Agency; and Man as a Moral Agent* (London, 1842).

MacCULLOCH, J., *A Geological Classification of Rocks, with Descriptive Synopses of the Species and Varieties, comprising the Elements of Practical Geology* (London, 1821).

—— *A System of Geology, with a Theory of the Earth, and an Explanation of its Connexion, with the Sacred Records* (2 vols., London, 1831).

—— *Proofs and Illustrations of the Attributes of God, from the Facts and Laws of the Physical Universe: being the Foundation of Natural and Revealed Religion* (3 vols., London, 1837).

McCULLOCH, J. R., *A Statistical Account of the British Empire* (2 vols., London, 1838).

—— *Principles of Political Economy* (Edinburgh, 1843).

—— *The Literature of Political Economy* (London, 1845).

MacENERY, J., *Cavern Researches, or, Discoveries of Organic Remains, and of British and Roman Reliques, in the Caves of Kent's Hole*, ed. E. Vivian (London, 1859).

McHENRY, J., *The Antediluvians, or, the World Destroyed; a Narrative Poem, in Ten Books* (London, 1839).

MACKENZIE, G., 'Description d'un tronc pétrifié hors de terre', *Bibliotheque universelle* viii (1818), 256–8.

MACLEAY, W. S., 'A Letter to J. E. Bicheno, Esq., F.R.S., in Examination of his Paper "On Systems and Methods"', in the Linnean Transactions', *Zool. Journ.* iv (1829), 401–15.

(MALTHUS, T. R.), *An Essay on the Principle of Population* (London, 1798).

MANTELL, G. A., *The Fossils of the South Downs; or Illustrations of the Geology of Sussex* (London, 1822).

—— 'Notice on the Iguanodon, a Newly Discovered Fossil Reptile, from the Sandstone of Tilgate Forest, in Sussex', *Phil. Trans.* cxv (1825), 179–86.

—— 'Remarks on the Geological Position of the Strata of Tilgate Forest in Sussex', *Edinburgh New Phil. Journ.* Apr. to Oct. 1826, 262–5.

—— *Illustrations of the Geology of Sussex: Containing a General View of the Geological Relations of the South-Eastern Part of England; with Figures and Descriptions of the Fossils of Tilgate Forest* (London, 1827).

—— 'The Geological Age of Reptiles', *Edinburgh New Phil. Journ.* Apr. to Oct. 1831, 181–5.

—— *The Geology of the South-east of England* (London, 1833).

(MANTELL, G. A.), *Thoughts on a Pebble; or a First Lesson in Geology* (London, 1836).

MANTELL, G. A., *The Wonders of Geology* (2 vols., London, 1838).

—— *The Medals of Creation; or, First Lessons in Geology, and in the Study of Organic Remains* (2 vols., London, 1844).

—— 'A few Notes on the Prices of Fossils', *London Geol. Journ.* i (1846), 13–17.

—— *Thoughts on Animalcules; or, a Glimpse of the Invisible World, Revealed by the Microscope* (London, 1846).

MARTIN, J., *A Descriptive Catalogue of the Engraving of the Deluge* (London, 1828).

(MILL, J. S.), 'Bentham: as a Jurist', *London and Westminster Review* vii (1838), 467–506.

MILLER, H., *The Old Red Sandstone; or, New Walks in an Old Field* (Edinburgh, 1841).

MILLER, J. S., *Prospectus for Publishing by Subscription, a Natural History of the Crinoeidea or Lily Shaped Zoophytes* (Bristol, 1821).

MILTON, J., *Paradise Lost,* ed. A. Fowler (London, 1971).

MITCHELL, T. L., 'An Account of the Limestone Caves at Wellington Valley, and of the Situation, near One of Them, where Fossil Bones have been Found', *Proc. Geol. Soc.* i (1834), 321–2.

MOYLE, M. P., 'On the Temperature of Mines', *Ann. Phil.* v (1823), 34–43; xi (1826), 259–60.

MOZLEY, T., *Reminiscences, chiefly of Oriel College and the Oxford Movement* (2 vols., London, 1882).

—— *Letter to the Rev. Canon Bull* (London, 1882).

MURCHISON, R. I., *Outline of the Geology of the Neighbourhood of Cheltenham* (Cheltenham, 1834).

—— 'The Gravel and Alluvia of S. Wales and Siluria', *Proc. Geol. Soc.* ii (1838), 230–6.

—— *The Silurian System, Founded on Geological Researches in the Counties of Salop, Hereford, Radnor, Montgomery, Caermarthen, Brecon, Pembroke, Monmouth, Gloucester, Worcester, and Stafford; with Descriptions of the Coal-fields and Overlying Formations* (2 vols., London, 1839).

—— Anniversary address to the Geological Society, 17 Feb. 1832, *Proc. Geol. Soc.* i (1834), 362–86.

—— Anniversary address to the Geological Society, 15 Feb. 1833, *Proc. Geol. Soc.* i (1834), 438–64.

—— Anniversary address to the Geological Society 18 Feb. 1842, *Proc. Geol. Soc.* iii (1842), 637–87.

—— Anniversary address to the Geological Society 17 Feb. 1843, *Proc. Geol. Soc.* iv (1844), 65–151.

MURCHISON, R. I., and E. DE VERNEUIL, 'On the Northern and Central Portions of Russia in Europe', *Proc. Geol. Soc.* iii (1842), 398–408.

NARES, E., *A View of the Evidences of Christianity at the Close of the Pretended Age of Reason* (Oxford, 1805).

—— *Man, as Known to Us Theologically and Geologically* (London, 1834).

(NEWMAN, J. H.), *'Memorials of Oxford'*, *British Critic* xxiv (1838), 133–46.

NEWMAN, J. H., *The Idea of a University*, ed. I. T. Kerr (Oxford, 1976).

NEWTON, I., *The Chronology of Ancient Kingdoms Amended* (London, 1728).

NIEUWENTYT, B., *Het Regt Gebruik der Werelt Beschouwingen, ter Overtuiginge van Ongodisten en Ongelovigen* (Amsterdam, 1715).

NÖGGENRATH, J., *Ueber aufrecht im Gebirgsgestein eingeschlossene fossile Baumstämme und andere Vegetabilien* (Bonn, 1819).

NOLAN, F., *The Time of the Millennium Investigated; and its Nature Determined on Scriptural Grounds* (London, 1831).

—— *The Analogy of Revelation and Science* (Oxford, 1833).

—— *The Chronological Prophecies; as Constituting a Connected System* (London, 1837).

OWEN, D. D., 'Regarding Human Foot-Prints in Solid Limestone', *Am. Journ. Sci.* xliii (1842), 14–32.

OWEN, R., 'Report on British Fossil Reptiles', *Report BAAS, 1839,* 13–126; *1841,* 60–204.

—— *Description of the Skeleton of an extinct Gigantic Sloth* (London, 1842).

—— 'Notice of the Glyptodon', *Proc. Geol. Soc.* iii (1842), 109–13.

—— 'Report on the British Fossil Mammalia', *Report BAAS, 1842,* 54–74.

—— *Geology and Inhabitants of the Ancient World* (London, 1854).

*The Works of William Paley, D.D.*, *Containing his Life, Moral and Political Philosophy, Evidences of Christianity, Natural Theology, Tracts, Horae Paulinae, Clergyman's Companion, and Sermons* (Edinburgh, 1825).

PANDER, C. H., and E.D'ALTON, *Das Riesen-Faultier, Bradypus giganteus, abgebildet, beschrieben, und mit den verwandten Geschlechten verglichen* (Bonn, 1821).

—— *Die vergleichende Osteologie*, i (Bonn, 1821–7).

PARISH, W., 'An Account of the Discovery of Portions of Three Skeletons of the Megatherium in the Province of Buenos Ayres in South America', *Proc. Geol. Soc.* i (1834), 403–4.

PARKINSON, J., *Organic Remains of a Former World. An Examination of the Mineralized Remains of the Vegetables and Animals of the Antediluvian World; Generally Termed Extraneous Fossils* (3 vols, London, 1804–11).

—— 'Observations on Some of the Strata in the Neighbourhood of London, and on the Fossil Remains in Them', *Trans. Geol. Soc.* i (1811), 324–54.

—— *Outlines of Oryctology. An Introduction to the Study of Fossil Organic Remains; especially those found in the British Strata* (London, 1822).

PEACOCK, G., 'Address', *Report BAAS, 1844*, xxxi–xlvi.

PENN, G., *A Christian's Survey of all the Primary Events and Periods of the World; from the Commencement of History to the Conclusion of Prophecy* (London, 1811).

—— *A Comparative Estimate of the Mineral and Mosaical Geologies* (London, 1822).

(PENN, G.), *Supplement to the Comparative Estimate of the Mineral and Mosaical Geologies: Relating Chiefly to the Geological Indications of the Phenomena of the Cave at Kirkdale* (London, 1823).

PERROT, A. M., *Tableau du monde antédiluvien, rédigé d'après Cuvier, Buckland, de Humboldt etc.* (Paris, undated).

PETAVIUS, D., *Opus de Doctrina Temporum* (2 vols., 1627–30).

—— *Rationarium Temporum* (Paris, 1652).

PHILLIPS, J., *Illustrations of the Geology of Yorkshire; or, a Description of the Strata and Organic Remains of the Yorkshire Coast: Accompanied by a Geological Map, Sections, and Plates of the Fossil Plants and Animals* (York, 1829).

—— *A Guide to Geology* (London, 1834).

—— *A Treatise on Geology* (2 vols., London, 1837).

—— *Figures and Descriptions of the Palaeozoic Fossils of Cornwall, Devon, and West Somerset* (London, 1841).

—— *Memoirs of William Smith* (London, 1844).

—— 'Letter to Cotton', *Replies to 'Essays and Reviews'* (Oxford, 1862), 514–16.

PHILLIPS, W., *A Selection of Facts from the Best Authorities Arranged so as to Form an Outline of the Geology of England and Wales* (London, 1818).

PIDGEON, E., *The Fossil Remains of the Animal Kingdom* (London, 1830).

PLAYFAIR, J., *Illustrations of the Huttonian Theory of the Earth* (Edinburgh, 1802).

(PLAYFAIR, J.), 'La Place, *Traité de Méchanique Céleste*', *Edinburgh Review* xi (1808), 249–84.

—— 'Woodhouse's *Trigonometry*', *Edinburgh Review* xvii (1810), 122–35.

—— 'Cuvier *on the Theory of the Earth*', *Edinburgh Review* xxii (1814), 454–75.

(PLAYFAIR, J.?), 'Dealtry's *Principles of Fluxions*', *Edinburgh Review* xxvii (1816), 87–98.

PONTOPPIDAN, E., *The Natural History of Norway* (London, 1755).

POWELL, B., *The Present State and Future Prospects of Mathematical and Physical Studies in the University of Oxford* (Oxford, 1832).

—— *Revelation and Science* (Oxford, 1833).

—— *The Connexion of Natural and Divine Truth; or the Study of the Inductive Philosophy considered as Subservient to Theology* (London, 1838).

PRESTWICH, J., 'On the Geology of Coalbrook Dale', *Trans. Geol. Soc.* v (1840), 413–96.

PRÉVOST, C., 'Observations sur les schistes oolithique de Stonesfield en Angleterre dans lesquelle ont trouvés plusieurs ossemens fossiles de mammifères', *Ann. sci. nat.* iv (1825), 389–417.

—— 'Les continens actuels ont-ils été, a plusieurs reprises, submergés par la mer?', *Mém. Soc. Hist. nat. Paris* iv (1828), 249–346.

PRICHARD, J. C., *An Analysis of Egyptian Mythology: to which is Subjoined, a Critical Examination of the Remains of Egyptian Chronology* (London, 1819).

—— *Researches into the Physical History of Mankind*, i (London, 1841).

QUENSTEDT, F. A., *Über Pterodactylus suevicus im Lithographischen Schiefer Württembergs* (Tübingen, 1855).

(RALEIGH, W.), *The History of the World* (London, 1614).

RAY, J., *Three Physico-Theological Discourses, Concerning (1) the Primitive Chaos and Creation of the World, (2) the General Deluge, its Causes and Effects, (3) the Dissolution of the World and Future Conflagration* (2nd edn., London, 1693).

READE, J. E., *The Deluge; a Drama in Twelve Scenes* (London, 1839).

(RENNIE, J.), *Conversations on Geology; Comprising a Familiar Explanation of the Huttonian and Wernerian Systems; the Mosaic Geology, as Explained by Mr. Granville Penn; and the Late Discoveries of Professor Buckland, Humboldt, Dr. Maculloch and Others* (London, 1840).

RHIND, W., *The Age of the Earth, Geologically and Historically* (Edinburgh, 1838).

RICHARDSON, G. F., *An Introduction to Geology, and its associate Sciences, Mineralogy, Fossil Botany, and Palaeontology* (London, 1851).

ROSENMÜLLER, J. C., *Beiträge zur Geschichte und nähern Kenntniss fossiler Knochen* (Leipzig, 1795).

*The Works of John Ruskin,* ii, *Poems,* ed. E. T. Cook and A. Wedderburn (London, 1903).

RUSSELL, M., *A Connection of Sacred and Profane History, from the Death of Joshua to the Decline of the Kingdoms of Israel and Judah* (3 vols., Edinburgh, 1827–37).

(SANFORD, D. K.), 'Classical Education', *Edinburgh Review* xxxv (1821), 302–14.

SAUSSURE, H.-B. DE, *Voyages dans les Alpes, précédés d'un essai sur l'histoire naturelle des environs de Genève,* i (Neuchâtel, 1780).

SCAFE, J., *The Genius, and other Poems* (Newcastle, 1819).

(SCAFE, J.), *King Coal's Levee, or Geological Etiquette, with Explanatory Notes; and the Council of the Metals* (London, 1820).

—— *Court News; or, the Peers of King Coal: and the Errants; or a Survey of the British Strata: with Explanatory Notes* (London, 1820).

—— *A Geological Primer in Verse: with a Poetical Geognosy, or Feasting and Fighting; and sundry right Pleasant Poems; with Notes. To which is Added a Critical Dissertation on 'King Coal's Levee', Addressed to the Professors and Students of the University of Oxford* (London, 1820).

SCALIGER, J. J., *Opus de Emendatione Temporum* (Geneva, 1729).

SCHEFFEL, J. V. VON, *Gaudeamus! Lieder aus dem Engeren und Weiteren* (Stuttgart, 1869).

SCHMERLING, P. C., *Recherches sur les ossemens fossiles découverts dans les cavernes de la province de Liège* (3 vols., Liège, 1833–4).

SCHOOLCRAFT, H. R., 'Remarks on the Prints of Human Feet, Observed in the Secondary Limestone of the Mississippi Valley', *Am. Journ. Sci.* v (1822), 223–31.

(SCROPE, G. P.), 'Lyell's *Principles of Geology*', *Quarterly Review* xliii (1830), 411–69; liii (1835), 406–48.

SCROPE, G. P., *Principles of Political Economy, Deduced from the Natural Laws of Social Welfare, and Applied to the Present State of Britain* (London, 1833).

—— 'On the Gradual Excavation of the Valleys in which the Meuse, the Moselle, and some other Rivers Flow', *Proc. Geol. Soc.* i (1834), 170–1.

(SCROPE, G.P.), 'Dr. Buckland's *Bridgewater Treatise*', *Quarterly Review* lvi (1836), 31–64.

SCROPE, G. P., *The Geology and Extinct Volcanos of Central France* (London, 1858).

SEDGWICK, A., 'On the Origin of Alluvial and Diluvial Formations', *Ann. Phil.* ix (1825), 241–57.

—— 'On Diluvial Formations', *Ann. Phil.* x (1825), 18–37.

—— *A Discourse on the Studies of the University* (Cambridge, 1833; 5th edn., 1850).

—— Anniversary address to the Geological Society, 19 Feb. 1830, *Proc. Geol. Soc.* i (1834), 187–212.

—— 'Address, on Announcing the First Award of the Wollaston Prize', *Proc. Geol. Soc.* i (1834), 270–9.

—— Anniversary address to the Geological Society, 18 Feb. 1831, *Proc. Geol. Soc.* i (1834), 281–316.

(SEDGWICK, A.), *'Natural History of Creation'*, *Edinburgh Review* lxxxii (1845), 1–85.

SEDGWICK, A., and R. I. MURCHISON, 'A Sketch of the Structure of the Eastern Alps; with Sections through the Newer Formations on the Northern Flanks of the Chain, and through the Tertiary Deposits of Styria, etc. etc.', *Trans. Geol. Soc.* iii (1835), 301–420.

SERRES, M. DE, 'Essai sur les cavernes à ossemens et sur les causes qui les y ont accumulés', *Natuurkundige Verhandelingen van de Hollandsche Maatschappij der Wetenschappen te Haarlem,* xxii (1835), 1–222.

(SEWELL, W.?), 'The British Association', *Oxford Univ. Mag.* i (1834), 401–12.

(SEWELL, W.), 'Animal Magnetism', *British Critic* xxiv (1838), 301–47.

*Shelley's 'Prometheus Unbound',* ed. L. J. Zillman (Seattle, 1959).

SHUCKFORD, S., *The Sacred and Profane History of the World Connected, from the Creation of the World to the Dissolution of the Assyrian Empire* (2 vols., London, 1728).

(SILLIMAN, B.), 'Geological Poems', *Am. Journ. Sci.* v (1822), 272–85.

—— 'Notice and Review of the Reliquiae Diluvianae', *Am. Journ. Sci.* viii (1824), 150–68, 317–38.

SMITH, A., *An Inquiry into the Nature and Causes of the Wealth of Nations* (London, 1776).

(SMITH, J. P.), 'Geology and the Holy Scriptures', *Mag. Pop. Sci.* ii (1836), 465–8.

SMITH, J. P., *The Relation between the Holy Scriptures and some Parts of Geological Science* (London, 1852).

SMITH, W., *A Delineation of the Strata of England and Wales, with Part of Scotland* (London, 1815).

—— *Stratigraphical System of Organized Fossils, with Reference to the Specimens of the Original Geological Collection in the British Museum: Explaining their State of Preservation and their Use in Identifying British Strata* (London, 1817).

SMITHSON, J., 'Some Observations on Mr. Penn's Theory Concerning the Formation of the Kirkdale Cave', *Ann. Phil.* viii (1824), 50–60.

STILLINGFLEET, E., *Origines Sacrae: or a Rational Account of the Grounds of Natural and Revealed Religion* (2 vols., Oxford, 1797).

STRANGWAYS, W. FOX-, 'Geological Sketch of the Environs of Petersburg', *Trans. Geol. Soc.* v (1821), 392–458.

STRICKLAND, H. E., 'On the Deposits of Transported Materials usually termed *Drift*, which Exist in the Counties of Worcester and Warwick', W. Jardine, *Memoirs of Huch Edwin Strickland* (London, 1858), 90–104.

—— 'On Geology, in Relation to the Studies of the University of Oxford', W. Jardine, *Memoirs of Hugh Edwin Strickland* (London, 1858), 207–22.

SUMNER, J. B., *A Treatise on the Records of the Creation, and on the Moral Attributes of the Creator; with particular Reference to the Jewish History, and to the Consistency of the Principle of Population with the Wisdom and Goodness of the Deity* (2 vols., London, 1833).

SUTCLIFFE, J., *A Short Introduction to the Study of Geology; Comprising a New Theory of the Elevation of the Mountains, and the Stratification of the Earth: in which the Mosaic Account of the Creation and the Deluge is Vindicated* (London, 1817).

—— *A Refutation of Prominent Errors in the Wernerian System of Geology; and in the Theories of other Writers* (London, 1819).

—— *The Geology of the Avon, being an Enquiry into the Order of the Strata and Mineral Productions of the District Washed by its Streams* (Bristol, 1822).

SWAN, J., *Speculum Mundi. Or a Glasse Representing the Face of the World; Shewing both that it did Begin, and Must also End: the Manner How, and the Time When, Being largely Examined* (Cambridge, 1635).

TAYLOR, W. C., *The Natural History of Society in the Barbarous and Civilized State: an Essay towards Discovering the Origin and Course of Human Improvement* (2 vols., London, 1840).

*The Poems of Tennyson,* ed. C. Ricks (London, 1969).

THOMPSON, J. V., 'Contributions towards the Natural History of the Dodo (Didus ineptus *Lin.*), a Bird which Appears to have become Extinct towards the End of the Seventeenth or Begining of the Eighteenth Century', *Mag. Nat. Hist.* ii (1829), 442–8.

THOMPSON, T., 'An Attempt to Ascertain the Animals Designated in the Scriptures by the Names Leviathan and Behemoth', *Mag. Nat. Hist.* viii (1835), 307–21.

(THOMPSON, T.), 'On the Animals Designated in the Scriptures by the Names of *Leviathan* and *Behemoth*', *Edinburgh New Phil. Journ.* xix (1835), 263–81.

TOURNAL, P., 'Human Bones found in the Caves of Bize, near Narbonne', *Edinburgh New Phil. Journ.* Oct. 1828 to Mar. 1829, 383–4.

TOWNSEND, J., *The Character of Moses Established for Veracity as an Historian. Recording Events from the Creation to the Deluge* (Bath, 1813).

TRIMMER, J., 'On the Diluvial Deposits of Caernarvonshire, between the Snowdon Chain of Hills and the Menai Strait, and on the Discovery of Marine Shells in Diluvial Sand and Gravel on the Summit of Moel Tryfane, near Caernarvon, 1000 ft. above the Level of the Sea', *Proc. Geol. Soc.* i (1834), 331–2.

TURNER, S., *The Sacred History of the World, as Displayed in the Creation and Subsequent Events to the Deluge* (London, 1832).

TURTON, T., *Natural Theology considered with reference to Lord Brougham's Discourse on that Subject* (Cambridge, 1836).

300    *Bibliography*

TYTLER, A. F., *Universal History, from the Creation to the Eighteenth Century*, ed. W. F. Tytler, i (London, 1834).

UNGER, F., *Die Urwelt in ihren verschiedenen Bildungsperioden* (Leipzig, 1847).

*An Universal History, from the Earliest Account of Time to the Present*, i (London, 1736).

URE, A., *A New System of Geology, in which the Great Revolutions of the Earth and Animated Nature, are Reconciled at once to Modern Science and Sacred History* (London, 1829).

USSHER, J., *Annales Veteris et Novi Testamenti, a Prima Mundi Origine Deducti* (Geneva, 1722).

VALENCIENNES, A., 'Observations upon the Fossil Jaws from the Oolitic Beds at Stonesfield, named *Didelphis Prevostii* and *Did. Bucklandii*', *Mag. Nat. Hist.* iii (1839), 1–10.

VOIGT, J. C. W., *Drey Briefe über die Gebirgslehre für Anfänger und Unkundige* (Weimar, 1786).

VOSSIUS, I., *Dissertatio de Vera Aetate Mundi* (The Hague, 1659).

WALLACE, R., *A Dissertation on the True Age of the World, in which is Determined the Chronology of the Period from Creation to the Christian Era* (London, 1844).

WEAVER, T., 'Geological Remarks', *Ann. Phil.* iv (1822), 81–98.

—— 'On Fossil Human Bones, and other Animal Remains Recently found in Germany', *Ann. Phil.* v (1823), 17–43.

WEBSTER, T., 'On the Freshwater Formations in the Isle of Wight, with some Observations on the Strata over the Chalk in the South-east Part of England', *Trans. Geol. Soc.* ii (1814), 161–254.

WEISSENBORN, W., 'On the Influence of Man in Modifying the Zoological Features of the Globe; with Statistical Accounts respecting a few of the more Important Species', *Mag. Nat. Hist.* ii (1838), 13–18.

WERNER, A. G., *Kurze Klassification und Beschreibung der verschiedenen Gebirgsarten* (Freiberg, 1787).

(WHATELY, R.), *Historic Doubts relative to Napoleon Buonaparte* (London, 1819).

WHATELY, R., *Introductory Lectures on Political Economy, being Part of a Course of Lectures delivered in Easter Term 1831* (London, 1831).

(WHATELY, R.), *Das Leben Napoleon's, kritisch geprüft; nebst einigen Nutzanwendungen auf 'Das Leben Jesu, von Strauss'* (Leipzig, 1836).

(WHEWELL, W.), 'Science of the English Universities', *British Critic* ix (1831), 71–90.

—— 'Lyell—*Principles of Geology*', *British Critic* ix (1831), 180–206.

WHEWELL, W., 'Progress of Geology', *Edinburgh New Phil. Journ.* Apr. to Oct. 1831, 242–7.

—— *Astronomy and General Physics Considered with Reference to Natural Theology* (London, 1833).

—— *History of the Inductive Sciences, from the Earliest to the Present Time*, iii (London, 1837).

—— Anniversary address to the Geological Society, 16 Feb. 1838, *Proc. Geol. Soc.* ii (1838), 624–49.

—— Anniversary address to the Geological Society, 15 Feb. 1839, *Proc. Geol. Soc.* iii (1842), 61–98.

—— *The Elements of Morality, including Polity* (2 vols., London, 1845).

—— *Indications of the Creator* (London, 1845).

(WHEWELL, W.), *Of the Plurality of Worlds: an Essay* (London, 1853).

WHITE, C., 'Observations on a Thigh Bone of Uncommon Length', *Mem. Lit. Phil. Soc. Manchester* ii (1785), 350–7.

WILCKE, H. C. D., 'Politia Naturae', C. Linnaeus, *Amoenitates Academicae; seu Dissertationes Variae Physicae, Medicae, Botanicae,* vi (Stockholm, 1763), 17–39.

WILLIAMS, I., *Ars Geologica: Poema Cancellarii Praemio Donatum* (Oxford, 1823).

WISEMAN, N. P. S., *Twelve Lectures on the Connexion between Science and Revealed Religion* (London, 1836).

WITHAM, H. T. M., *Observations on Fossil Vegetables accompanied by Representations of their Internal Structure, as seen through the Microscope* (Edinburgh, 1831).

—— *The Internal Structure of Fossil Vegetables found in the Carboniferous and Oolitic Deposits of Great Britain* (Edinburgh, 1833).

YOUNG, G., 'Account of a Fossil Crocodile recently Discovered in the Alum-Shale near Whitby', *Edinburgh Phil. Journ.* xiii (1825), 76–81.

—— 'On the Fossil Remains of Quadrupeds, etc. Discovered in the Cavern at Kirkdale, in Yorkshire, and in other Cavities or Seams in Limestone Rocks', *Mem. Wernerian Nat. Hist. Soc.* iv (1822), 262–70; vi (1832), 171–83.

YOUNG, G., and J. BIRD, *A Geological Survey of the Yorkshire Coast: describing the Strata and Fossils occurring between the Humber and the Tees, from the German Ocean to the Plain of York* (Whitby, 1828).

SECONDARY LITERATURE

ABBOTT, E., and L. CAMPBELL, *The Life and Letters of Benjamin Jowett* (2 vols., London, 1897).

ADAMS, E., *Francis Danby: Varieties of Poetic Landscape* (New Haven, 1973).

AGASSIZ, E. C., *Louis Agassiz. His Life and Correspondence* (2 vols., London, 1885).

ALBRITTON, C. C., *The Abyss of Time. Changing Conceptions of the Earth's Antiquity after the Sixteenth Century* (San Francisco, 1980).

ALLEN, D. E., *The Naturalist in Britain. A Social History* (London, 1976).

*Archives of the Royal Institution of Great Britain; Minutes of Managers' Meetings, 1799–1900,* v, introduced by M. Berman (Ilkley, 1975).

ARKELL, W. J., *The Jurassic System in Great Britain* (Oxford, 1933).

BAILEY, E. B., *James Hutton, the Founder of Modern Geology* (Amsterdam, 1967).

BARBER, L., *The Heyday of Natural History* (London, 1980).

BARR, J., *Fundamentalism* (London, 1977).

BARTHOLOMEW, M., 'The Non-Progress of Non-Progression: Two Responses to Lyell's Doctrine', *Brit. Journ. Hist. Sci.* ix (1976), 166–74.

BAUMGÄRTEL, H., 'Alexander von Humboldt: Remarks on the Meaning of Hypothesis in his Geological Researches', *Toward a History of Geology,* ed. C. J. Schneer (Cambridge, Mass., 1969), 19–35.

BERMAN, M., *Social Change and Scientific Organisation; the Royal Institution, 1799–1844* (London, 1978).

BOASE, G. C., and W. P. COURTNEY, *Bibliotheca Cornubiensis,* ii (London, 1878).

BOURDIER, F., 'Geoffroy Saint-Hilaire versus Cuvier: the Campaign for Paleontological Evolution (1825–1838)', *Toward a History of Geology,* ed. C. J. Schneer (Cambridge, Mass., 1969), 36–61.

BOWLER, P. J., *Fossils and Progress; Palaeontology and the Idea of Progressive Evolution in the Nineteenth Century* (New York, 1976).

BOYLAN, P. J., 'Dean William Buckland, 1784–1856. A Pioneer in Cave Science', *Studies in Speleology* i (1967), 237–53.

BRIGGS, A., *The Age of Improvement 1783–1867* (London, 1979).

BROCK, M., *The Great Reform Act* (London, 1973).

BROOKE, J. H., 'Natural Theology and the Plurality of Worlds: Observations on the Brewster-Whewell Debate', *Ann. Sci.* xxxiv (1977), 221–86.

—— 'Richard Owen, William Whewell, and the *Vestiges*', *Brit. Journ. Hist. Sci.* x (1977), 132–45.

—— 'The Natural Theology of the Geologists: some Theological Strata', *Images of the Earth,* ed. L. J. Jordanova and R. S. Porter (Chalfont St. Giles, Bucks., 1979), 39–64.

DE BRUIJN, J. G., *Inventaris van de Prijsvragen uitgeschreven door de Hollandsche Maatschappij der Wetenschappen, 1753–1917* (Groningen, 1977).

BUCKLAND, F. T., 'Memoir of the Very Rev. William Buckland', W. Buckland, *Geology and Mineralogy,* i (3rd edn., London, 1858), xvii–lxx.

—— *Curiosities of Natural History,* 1st and 2nd series (London, 1893).

BURCHFIELD, J. D., *Lord Kelvin and the Age of the Earth* (London, 1975).

BURY, J. B., *The Idea of Progress. An Inquiry into its Origin and Growth* (London, 1920).

*Byron's Letters and Journals,* ed. L. A. Marchand, viii (London, 1978).

CANNON, S. F., *Science in Culture: the Early Victorian Period* (New York, 1978).

CASIER, E., *Les Iguanodons de Bernissart* (Brussels, 1960).

CHADWICK, O., *The Victorian Church,* i (London, 1971).

CHORLEY, R. J., A. J. DUNN, and R. P. BECKINSALE, *The History of the Study of Landforms or the Development of Geomorphology,* i (London, 1964).

CHURCH, R. W., *The Oxford Movement; Twelve Years, 1833–1845,* ed. and introduced by G. Best (Chicago, 1970).
CLARK, J. W., and T. M. HUGHES, *The Life and Letters of the Reverend Adam Sedgwick* (2 vols., Cambridge, 1890).
COHN, N., *The Pursuit of the Millennium* (London, 1957).
COLLINGWOOD, R. G., *The Idea of History* (London, 1946).
CORSI, P., 'The Importance of French Transformist Ideas for the Second Volume of Lyell's *Principles of Geology'*, *Brit. Journ. Hist. Sci.* xi (1978), 221–44.
COX, G. V., *Recollections of Oxford* (London, 1870).
CULLER, A. D., *The Poetry of Tennyson* (New Haven, 1977).
CUMMING, D. A., 'John MacCulloch, F.R.S., at Addiscombe; the Lectureships in Chemistry and Geology', *Notes and Records Roy. Soc.* xxxiv (1980), 155–83.
CUNNINGHAM, F. F., *The Revolution in Landscape Science* (Vancouver, 1977).
DAVIES, G. L., *The Earth in Decay. A History of British Geomorphology, 1578–1878* (London, 1969).
DEACON, M., *Scientists and the Sea, 1650–1900; a Study of Marine Science* (London, 1971).
DELAIR, J. B., 'A History of the Early Discoveries of Liassic Ichthyosaurs in Dorset and Somerset (1779–1835)', *Proc. Dorset Nat. Hist. Archaeol. Soc.* xc (1968), 115–27.
DELAIR, J. B., and W. A. S. SARJEANT, 'The Earliest Discoveries of Dinosaurs', *Isis* lxvi (1975), 5–25.
—— 'Joseph Pentland—a Forgotten Pioneer in the Osteology of Fossil Marine Reptiles', *Proc. Dorset Nat. Hist. Archaeol. Soc.* xcvi (1975), 12–16.
DOUGLAS, S., *The Life and Selections from the Correspondence of William Whewell* (London, 1881).
EDMONDS, J. M., 'The Founding of the Oxford Readership in Geology, 1818', *Notes and Records Roy. Soc.* xxxiv (1979), 33–51.
EDMONDS, J. M., and J. A. DOUGLAS, 'William Buckland, F.R.S. (1784–1856) and an Oxford Geological Lecture, 1823', *Notes and Records Roy. Soc.* xxx (1976), 141–67.
*The George Eliot Letters,* viii, ed. G. S. Haight (New Haven, 1978).
ENGEL, A., 'The Emerging Concept of the Academic Profession at Oxford', *The University in Society,* ed. L. Stone (Princeton, 1975), i. 305–52.
—— 'From Clergyman to Don: the Rise of the Academic Profession in Nineteenth Century Oxford', Ph.D. thesis, Princeton, 1975. (OUP book forthcoming).
ENGELHARDT, D. VON, *Historisches Bewusstsein in der Naturwissenschaft von der Aufklärung bis zum Positivismus* (München, 1979).
ENGELHARDT, W. VON, 'Die Entwicklung der geologischen Ideen seit

der Goethe-Zeit', *Abh. Braunschweigischen Wissenschaftlichen Gesellschaft,* 1979, 1–23.

—— 'Neptunismus und Plutonismus', *Fortschritte der Mineralogie* lx (1982), 21–43.

EYLES, J. M., 'G. W. Featherstonhaugh (1780–1866), F.R.S., F.G.S., Geologist and Traveller', *Journ. Soc. Biblphy Nat. Hist.* viii (1978), 381–95.

FABER, G., *Oxford Apostles; a Character Study of the Oxford Movement* (London, 1936).

FEAVER, W., *The Art of John Martin* (Oxford, 1975).

FISHER, G. P., *Life of Benjamin Silliman* (2 vols., New York, 1866).

FOUCAULT, M., *Les Mots et les choses. Une archéologie des sciences humaines* (Paris, 1966).

GARLAND, M. M., *Cambridge before Darwin; the Ideal of a Liberal Education, 1800–1860* (Cambridge, 1980).

GEIKIE, A., *The Scottish School of Geology* (Edinburgh, 1871).

—— *Life of Sir Roderick I. Murchison Based on his Journals and Letters* (2 vols., London, 1875).

—— *The Founders of Geology* (Dover Reprint, New York, 1962).

GILLISPIE, C. C., *Genesis and Geology. A Study in the Relations of Scientific Thought, Natural Theology, and Social Opinion in Great Britain, 1790–1850* (New York, 1959).

—— *The Edge of Objectivity* (Princeton, 1960).

GLASS, B., O. TEMKIN, and W. L. STRAUS (eds.), *Forerunners of Darwin: 1745–1859* (Baltimore, 1959).

*Goethe. Die Schriften zur Naturwissenschaft,* ed. K. L. Wolf, W. Troll, R. Matthaei, W. von Engelhardt, and D. Kuhn, xi (Weimar, 1970).

GORDON, E. O., *The Life and Correspondence of William Buckland* (London, 1894).

GOTTSCHALK, C. G., 'Verzeichniss Derer, welche seit Eröffnung der Bergakademie und bis Schluss des ersten Säculum's auf ihr studirt haben', *Festschrift zum hundertjährigen Jubiläum der Königl. Sächs. Bergakademie zu Freiberg* (Dresden, 1866), 221–95.

GOULD, S. J., 'Is Uniformitarianism Necessary?', *Am. Journ. Sci.* cclxiii (1965), 223–8.

—— *Ontogeny and Phylogeny* (Cambridge, Mass., 1977).

GRAFTON, A. T., 'Joseph Scaliger and Historical Chronology: the Rise and Fall of a Discipline', *History and Theory* xiv (1975), 156–85.

GREEN, V. H. H., *Religion at Oxford and Cambridge* (London, 1964).

GREENACRE, F., *The Bristol School of Artists. Francis Danby and Painting in Bristol, 1810–1840* (Bristol, 1973).

HABER, F. C., *The Age of the World. Moses to Darwin* (Baltimore, 1959).

—— 'Fossils and the Idea of a Process of Time in Natural History',

*Forerunners of Darwin: 1745–1859*, ed. B. Glass, O. Temkin, and W. L. Straus (Baltimore, 1959), 222–61.

HACHISUKA, M., *The Dodo and Kindred Birds or the Extinct Birds of the Mascarene Islands* (London, 1953).

HANSEN, B., 'The Early History of Glacial Theory in British Geology', *Journ. Glaciology* ix (1970), 135–41.

HARRISON, J. F. C., *The Second Coming. Popular Millenarianism 1780–1850* (London, 1979).

HELLER, F., 'Die Forschungen in der Zoolithenhöhle bei Burggaillenreuth von Esper bis zur Gegenwart', *Erlanger Forschungen* v (1972), 7–56.

HEUVELMANS, B., *In the Wake of the Sea-Serpents* (New York, 1968).

HILL, C., *Antichrist in Seventeenth-Century England* (London, 1971).

HOOYKAAS, R., *Natural Law and Divine Miracle. The Principle of Uniformity in Geology, Biology and Theology* (Leiden, 1963).

HOWORTH, H. H., 'The Origin and Progress of the Modern Theory of the Antiquity of Man', *Geol. Mag.* ix (1902), 16–27.

HUNT, A. R., 'On Kent's Cavern with Reference to Buckland and his Detractors', *Geol. Mag.* ix (1902), 114–18.

HUNT, J., *Religious Thought in England in the Nineteenth Century* (London, 1896).

JAHN, M. E., 'Some Notes on Dr. Scheuchzer and on *Homo diluvii testis*', *Toward a History of Geology*, ed. C. J. Schneer (Cambridge, Mass., 1969), 193–213.

JARDINE, W., *Memoirs of Hugh Edwin Strickland* (London, 1858).

KENNARD, A. S., 'The Early Digs in Kent's Hole, Torquay, and Mrs. Cazalet', *Proc. Geol. Assoc.* lvi (1945), 156–213.

KRÄMER, F., and H. KUNZ, '*Chirotherium,* das ''unbekannte'' Tier', *Natur und Museum* xcvi (1966), 12–19.

LAUDAN, R., 'Ideas and Organisation in British Geology: a Case Study in Institutional History', *Isis* lxviii (1977), 527–38.

LAWRENCE, P., 'Heaven and Earth—the Relation of the Nebular Hypothesis to Geology', *Cosmology, History, and Theology*, ed. W. Yourgrau and A. D. Breck (New York, 1977), 253–81.

—— 'Charles Lyell versus the Theory of Central Heat: a Reappraisal of Lyell's Place in the History of Geology', *Journ. Hist. Biol.* xi (1978), 101–28.

LELAND, C. G., *Gaudeamus! Translated from the German* (London, 1872).

LEPENIES, W., *Das Ende der Naturgeschichte; Wandel kultureller Selbstverständlichkeiten in den Wissenschaften des 18. und 19. Jahrhunderts* (München, 1976).

LOVEJOY, A. O., *The Great Chain of Being. A Study of the History of an Idea* (New York, 1960).

LUCAS, J. R., 'Wilberforce and Huxley: a Legendary Encounter', *Hist. Journ.* xxii (1979), 313–30.

306     *Bibliography*

LURIE, E., *Louis Agassiz. A Life in Science* (Chicago, 1960).
LYELL, K. M., *Life, Letters, and Journals of Sir Charles Lyell* (2 vols., London, 1881).
LYON, J., 'The Search for Fossil Man: Cinq Personnages à la Recherche du Temps Perdu', *Isis* lxi (1970), 68–84.
McCARTNEY, P. J., *Henry De la Beche: Observations on an Observer* (Cardiff, 1977).
McKERROW, R. E., 'Richard Whately on the Nature of Human Knowledge in Relation to Ideas of his Contemporaries' *Journ. Hist. Ideas* xlii (1981), 439–55.
*The Journal of Gideon Mantell, Surgeon and Geologist, covering the Years 1818–1852,* ed. and introduced by E. C. Curwen (London, 1940).
MATHIAS, P., *The First Industrial Nation; an Economic History of Britain, 1700–1914* (London, 1969).
MERZ, J. T., *A History of European Thought in the Nineteenth Century,* i (Edinburgh, 1896).
MEYER, H., 'The Age of the World. A Chapter in the History of Enlightenment', multigraphed essay (Allentown, Penna, 1951).
MILLHAUSER, M., 'The Scriptural Geologists. An Episode in the History of Opinion', *Osiris* xi (1954), 65–86.
—— *Just before Darwin: Robert Chambers and Vestiges* (Middletown, Conn., 1959).
MORRELL, J. B., 'The University of Edinburgh in the Late Eighteenth Century: its Scientific Eminence and Academic Structure', *Isis* lxii (1971), 158–71.
MORRELL, J., and A. THACKRAY, *Gentlemen of Science; Early Years of the British Association for the Advancement of Science* (Oxford, 1981).
MORRIS, J., *The Oxford Book of Oxford* (Oxford, 1978).
NEVE, M., and R. PORTER, 'Alexander Catcott: Glory and Geology', *Brit. Journ. Hist. Sci.* x (1977), 37–60.
*The Letters and Diaries of John Henry Newman,* ed. I. Ker and T. Gornall, i (Oxford, 1978).
NISBET, R., *History of the Idea of Progress* (London, 1980).
NORTH, F. J. 'Dean Conybeare, Geologist', *Report and Trans. Cardiff Naturalists' Soc.* lxvi (1933, published 1935), 15–68.
—— 'Paviland Cave, the "Red Lady", the Deluge, and William Buckland', *Ann. Sci.* v (1942), 91–128.
NORTH, J. D., 'Chronology and the Age of the World', *Cosmology, History, and Theology,* ed. W. Yourgrau and A. D. Breck (New York, 1977), 307–33.
OLDROYD, D. R., 'Historicism and the Rise of Historical Geology, Part i', *Hist. Sci.* xvii (1979), 191–213.
ORANGE, A. D., 'Hyaenas in Yorkshire: William Buckland and the Cave at Kirkdale', *History Today* xxii (1972), 777–85.

OSPOVAT, A. M., 'Reflections on A. G. Werner's "Kurze Klassifikation"', *Toward a History of Geology*, ed. C. J. Schneer (Cambridge, Mass., 1969), 242–56.

OSPOVAT, D., 'Lyell's Theory of Climate', *Journ. Hist. Biol.* x (1977), 317–39.

—— *The Development of Darwin's Theory; Natural History, Natural Theology, and Natural Selection, 1838–1859* (Cambridge, 1981).

PAGE, L. E., 'Diluvialism and its Critics in Great Britain in the Early Nineteenth Century', *Toward a History of Geology*, ed. C. J. Schneer (Cambridge, Mass., 1969), 257–71.

PENDERED, M. L., *John Martin, Painter. His Life and Times* (London, 1923).

PFEIFFER, R., *History of Classical Scholarship from 1300 to 1850* (Oxford, 1976).

POINTON, M. R., *Milton and English Art* (Manchester, 1970).

—— 'Geology and Landscape Painting in Nineteenth-Century England', *Images of the Earth*, ed. L. J. Jordanova and R. S. Porter (Chalfont St. Giles, 1979), 84–108.

POLLARD, S., *The Idea of Progress* (London, 1968).

PORTER, R. *The Making of Geology, Earth Science in Britain, 1660–1815* (Cambridge, 1977).

—— 'Creation and Credence: the Career of Theories of the Earth in Britain, 1660–1820', *Natural Order; Historical Studies of Scientific Culture*, ed. B. Barnes and S. Shapin (London, 1979), 97–123.

PRICE, G. MCCREADY, *Evolutionary Geology and the New Catastrophism* (Mountain View, Calif., 1926).

RASHID, S., 'Richard Whately and Christian Political Economy at Oxford and Dublin', *Journ. Hist. Ideas* xxxviii (1977), 147–55.

RAVEN, C. E., *Natural Religion and Christian Theology* (Cambridge, 1953).

REEVES, M., *The Influence of Prophecy in the Later Middle Ages* (Oxford, 1969).

RUDWICK, M. J. S., 'The Foundation of the Geological Society of London: its Scheme for Co-operative Research and its Struggle for Independence', *Brit. Journ. Hist. Sci.* i (1963), 325–55.

—— 'Lyell on Etna, and the Antiquity of the Earth', *Toward a History of Geology*, ed. C. J. Schneer (Cambridge, Mass., 1969), 288–304.

—— 'The Strategy of Lyell's *Principles of Geology*', *Isis* lxi (1970), 4–33.

—— 'Uniformity and Progression: Reflections on the Structure of Geological Theory in the Age of Lyell', *Perspectives in the History of Science and Technology*, ed. D. H. D. Roller (Norman, Oklahoma, 1971), 209–27.

—— *The Meaning of Fossils; Episodes in the History of Palaeontology* (London, 1972).

—— 'Poulett Scrope on the Volcanoes of Auvergne: Lyellian Time and Political Economy', *Brit. Journ. Hist. Sci.* vii (1974), 205–42.

—— 'Caricature as a Source for the History of Science: de la Beche's anti-Lyellian Sketches of 1831', *Isis* lxvi (1975), 534–60.

—— 'Historical Analogies in the Geological Work of Charles Lyell', *Janus* lxiv (1977), 89–107.

RUSCH, W. H., 'Human Footprints in Rocks', *Creation Research Society Quarterly*, Mar. 1971, 201–13.

*The Ruskin Family Letters*, ed. V. A. Bird, ii (Ithaca, 1973).

SARJEANT, W. A. S., 'A History and Bibliography of the Study of Vertebrate Footprints in the British Isles', *Palaeogeogr. Palaeoclim. Palaeoecol.* xvi (1974), 265–378.

SARJEANT, W. A. S., and J. B. DELAIR, 'An Irish Naturalist in Cuvier's Laboratory; the Letters of Joseph Pentland, 1820–1832', *Bull. Brit. Mus. (Nat. Hist.)*, Hist. Ser. vi (1980), 245–319.

SCHENK, H. G. A. V., *The Mind of the European Romantics* (London, 1966).

SCHULZ, G., 'Novalis und der Bergbau', *Freiberger Forschungshefte* D 11 (1955), 242–55.

SEZNEC, J., *John Martin en France* (London, 1964).

SHEPPARD, T., 'William Smith: his Maps and Memoirs', *Proc. Yorkshire Geol. Soc.* xix (1917), 73–253.

SIEGFRIED, R., 'Davy's "Intellectual Delight" and his Lectures at the Royal Institution', *Science and the Sons of Genius: Studies on Humphry Davy*, ed. S. Forgan (London, 1980), 177–99.

STANLEY, A. P., *The Life and Correspondence of Thomas Arnold* (London, 1844).

STEFFAN, T. G., *Byron's 'Don Juan'*, i (Austin, 1957).

—— *Lord Byron's 'Cain'* (Austin, 1968).

STEVENSON, L., *Darwin among the Poets* (Chicago, 1932).

*Tennyson. In Memoriam*, ed. J. D. Hunt (London, 1970).

TENNYSON, H., *Alfred Lord Tennyson. A Memoir* (2 vols., London, 1897).

THACKRAY, J. C., 'James Parkinson's *Organic Remains of a Former World* (1804–1811)', *Journ. Soc. Biblphy Nat. Hist.* vii (1976), 451–66.

TODHUNTER, I., *William Whewell* (2 vols., London, 1876).

TOON, P., *Puritans, the Millenium and the Future of Israel* (Cambridge, 1970).

TUCKWELL, W., *Reminiscences of Oxford* (London, 1907).

TURNER, P., *Tennyson* (London, 1976).

WAGENBRETH, O., 'Abraham Gottlob Werner und der Höhepunkt des Neptunistenstreites um 1790', *Freiberger Forschungshefte* D 11 (1955), 183–241.

—— 'Abraham Gottlob Werners System der Geologie, Petrographie und Lagerstättenlehre', *Freiberger Forschungshefte* C 223 (1967), 83–148.

—— 'Werner-Schüler als Geologen und Bergleute und ihre Bedeutung für die Geologie und Bergbau des 19. Jahrhunderts', *Freiberger Forschungshefte* C 223 (1967), 163–78.

WARD, W. R., *Victorian Oxford* (London, 1965).

WEBSTER, C., *The Great Instauration; Science, Medicine and Reform, 1626–1660* (London, 1975).

WEINDLING, P. J., 'Geological Controversy and its Historiography: the Prehistory of the Geological Society of London', *Images of the Earth,* ed. L. J. Jordanova and R. S. Porter (Chalfont St. Giles, 1979), 248–71.

WENDT, H., *Before the Deluge* (London, 1968).

WHATELY, E. J., *Life and Correspondence of Richard Whately, D.D., late Archbishop of Dublin* (2 vols., London, 1866).

WHITCOMB, J. C., and H. M. MORRIS, *The Genesis Flood. The Biblical Record and its Scientific Implications* (Philadelphia, 1962).

WHITE, G. W., 'Announcement of Glaciation in Scotland; William Buckland (1784–1856)', *Journ. Glaciology* ix (1970), 143–5.

WILLSHER, A. P., 'Daubeny and the Development of the Chemistry School in Oxford, 1822–1867', BA thesis (Oxford, 1961).

WILSON, L. G., *Charles Lyell: the Years to 1841: the Revolution in Geology* (New Haven, 1972).

WINSTANLEY, D. A., *Early Victorian Cambridge* (Cambridge, 1940).

WOODWARD, H. B., 'Dr. Buckland and the Glacial Theory', *Midland Naturalist* vi (1883), 225–9.

—— *The History of the Geological Society* (London, 1908).

YEO, R., 'William Whewell, Natural Theology and the Philosophy of Science in Mid Nineteenth Century Britain', *Ann. Sci.* xxxvi (1979), 493–516.

YOUNG, M. B., *Richard Wilton, a Forgotten Victorian* (London, 1967).

YULE, J. D., 'The Impact of Science on British Religious Thought in the Second Quarter of the Nineteenth Century', Ph.D. thesis (Cambridge, 1976).


*Reference works*

*Allgemeine Deutsche Biographie* (45 vols., Leipzig, 1875–1900)

*Dictionary of Anonymous and Pseudonymous English Literature,* ed. J. Kennedy, W. A. Smith, and A. F. Johnson (7 vols., Edinburgh, 1926–37).

*Dictionary of the History of Ideas,* ed. P. P. Wiener (5 vols., New York, 1968–74)

*Dictionary of National Biography* (21 vols., London, 1908–9)

*Dictionary of Scientific Biography,* ed. C. C. Gillispie (15 vols., New York, 1970–8)

*The Encyclopaedia Britannica* (11th edn., 29 vols., Cambridge, 1910–11)

*Nouvelle Biographie générale* (46 vols., Paris, 1855–66)

*Die Religion in Geschichte und Gegenwart* (3rd edn., 7 vols., Tübingen, 1957–65)

*The Wellesley Index to Victorian Periodicals, 1824–1900,* ed. W. E. Houghton (3 vols., Toronto, 1966–79)

# Index

Proper names include those of authors of articles and books cited in the footnotes, but generally not editors of multi-author volumes, nor writers of unpublished correspondence.